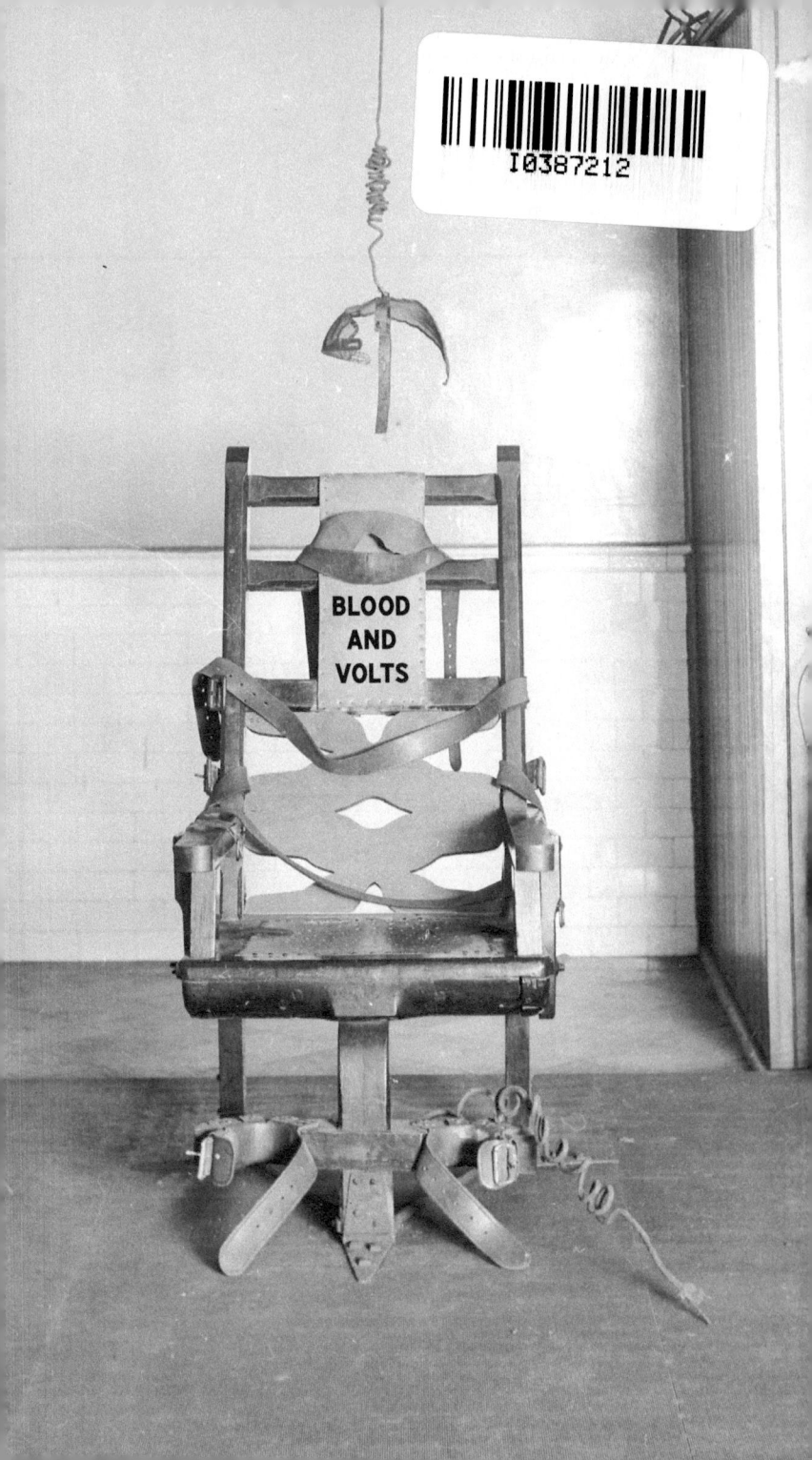

Execution of Ruth Snyder, 1928. Surreptitiosly photographed by a *New York Daily News* reporter with a camera strapped to his leg.

BLOOD AND VOLTS

EDISON, TESLA, AND THE ELECTRIC CHAIR

TH. METZGER

UNDERWORLD AMUSEMENTS

Blood and Volts: Edison, Tesla, and the Electric Chair
copyright ©1996, 2023 Thom Metzger.
www.ThomMetzger.com

First edition Autonomedia, 1996.
Second edition Underworld Amusements, 2024.

Cover and interior designed and typeset by Kevin I. Slaughter

Published March, 2024.

Hardback (ltd. ed.) isbn: 978-1-943687-16-9
Paperback isbn: 978-1-943687-31-2
eBook isbn: 978-1-943687-32-9

Underworld Amusements
Baltimore, MD
www.UnderworldAmusements.com

Contents

INTRODUCTION: **The Body in the Chair** 7
1. **The Death Commission** 19
2. **The Ark of the Ax** 55
3. **The Wizard** 69
4. **The Transformer** 99
5. **The War of the Currents** 133
6. **Cruel and Unusual** 167
7. **The Execution** 199
8. **Aftermath** 231
9. **Apotheosis** 251
BIBLIOGRAPHY 263
NAME INDEX 267

Awaiting execution, September 12, 1908, unidentified site.
(Photograph courtesy Library of Congress.)

Introduction

THE BODY IN THE CHAIR

The chair is a blunt, crude object. It was made by prison labor from solid oak: hard, heavy, and unyielding. It has only three legs, two in back and one broader in the front, with a crosspiece for the ankle straps. The chair's back is high, and at the top are two vertical slats that act as a headrest. The chair is bolted to the floor, solid and immovable as the high altar in a cathedral.

Jesse Joseph Tafero was strapped in securely. A wide leather belt across his lower chest, a left and right strap binding his wrists to the armrests, and a broader one across the top of his thighs like a primitive car seat belt. His feet too where held in place, within the T-shaped stocks at the base of the chair's wider third leg.

He was brought to the death chamber at Florida's Starke State Prison on the morning of May 4, 1990, one hundred years from the date of the electric chair's first use.

Like part of a homemade bondage outfit, the head piece was brought down over his face and cinched in tightly. Hard, crude leather, with no opening but a breather-slit for his nose. Tafero struggled, but not long. There was no point now. The only meaningful resistance in this room was measured in ohms. Tafero waited—blind and motionless—as the electrician slid the opening of the lead-line onto the contact post on the ground-pad. A few drops of salt water fell from Tafero's chin, having run down his face from the brine-soaked contact sponge in the headpiece.

The superintendent gave the signal and the executioner pushed the first electrical transfer button at the top of the circuit breaker. There came a dull mechanical hum as power was routed to the death switch on the control panel. Tafero stirred against the straps, wriggled and moaned. But there was a brief respite. The witnesses stared, holding their breath, clutching the seats of the cheap metal folding chairs. Silently, a light came on over the red death switch, telling everyone there—except Tafero—that all was ready. A low, hoarse monosyllable came from the superintendent's lips. "Now." The executioner moved the death switch one half turn to the left.

As the current surged through him, Tafero jolted upward in a violent total-body spasm. Immediately, the witnesses smelled the cloying sweet odor of burnt human flesh. Loud and continuous, the sound of frying bacon filled the room. Then twelve-inch blue and orange spikes of flame shot from the sides of Tafero's head. He nodded and gurgled, heaved against the straps, for four minutes, as the current ran from his head to the lower contact on his leg. Witnesses stood stupefied, nauseated. The flames hissed and flickered. Tafero's eyebrows burned off and fell to his shoulders like filthy snow.

The superintendent signaled for the current to be shut off. For a moment, witnesses, officials, technicians, and guards stood silently. Then a faint noise came from Tafero's mouth. He was still alive, gasping for breath, the sound muffled by the heavy leather mask.

Again came the signal for the current to be turned on. Again Tafero bucked up against the belts, his groans now drowned out by the chair's mechanical hissing. The spikes of flame shot out again, further now, accompanied by jets of bitter smoke.

For fear that the leather bindings would catch fire, the superintendent had the power cut a second time. Tafero's skin was bright red, blisters rising on the backs of his hands. Smoke swirled around his head like a dismal halo. But still he wasn't dead.

A third time the power was shunted into Tafero's head. As the temperature rose, his flesh swelled and the skin stretched near to the breaking point. He fought on, trapped, blind, helpless. Now a boiling acrid cloud rose from the seat, his urine turning to steam. He twitched and squirmed like an insect.

And again the power was cut. Witnesses leaned forward, listening, peering into the murky air.

"Oh, Jesus, he is still alive."

The superintendent slashed the air with his fist. "Now!"

The current blasted into Tafero and he thrashed impotently against the discharge. A last jolt of high voltage power, the last throes of pain, and it was over. The switch was turned back to the off position and the overloaded sizzling sound finally died.

Afterward, when the body was cool enough to remove, officials found that in the upper contact—the death cap—the natural elephant's-ear-sponge had been replaced by a common synthetic sponge. In earlier executions, a bare metal contact was found to dry out the skin and cause burning of the flesh, so a sponge soaked in a highly conductive brine solution was introduced to make better contact. A maintenance man had replaced the natural sponge with one of the common household type, which dried out quickly and reduce the voltage from the usual 2,000 to as low as 100. He'd apparently tested the flammability of the sponge by using a household toaster, and thought it sufficient.

The autopsy was delayed, as medical officials waited for the body—blistered, blackened and charred—to cool enough to safely touch.

State prison medical director Frank Kligo, who was present at the execution, stated later, "It was less than aesthetically attractive."[1]

The first humans killed by electricity were those rare individuals struck by lightning. In many prescientific cultures, places and people hit by lightning were thought to be transformed, imbued with unearthly power. Lightning was believed to be supernatural and sacred, the ultimate weapon of the gods. Later, more naturalistic

---

[1] Jacob Weisberg, "This Is Your Death," *New Republic* 1 July 1991: 23-27.

explanations were offered, one of the most persistent being that lightning was an explosion of so-called sulfurous vapors produced by marshes. Not until the 1700s was the connection between lightning and electricity made. And it was not until well into the 1800s that the nature of lightning—a discharge of atmospheric static electricity—was truly understood.

In 1832 Michael Faraday discovered the principle of induction—using magnetism to produce electricity—and within a year the first alternating current generator was created. Quickly, new uses for electricity, such as the telegraph and electric motor, were available, beginning the rapid embracing of electricity by Western culture.

Electrical current passing through the human body can cause death in a number of ways. In some, it paralyzes the muscular functioning of the heart. In others, it causes ventricular fibrillation (high-speed, uncontrolled twitching of the cardiac muscle). And it sometimes shuts down the breathing center of the brain. In addition to these effects, electricity can cause death by overheating the body. Temperatures of well over 200°F are commonly reported. According to Fred A. Leutcher, the president of the only company in America that designs and builds execution systems, "Current cooks, so it's important to limit the current. If you overload an individual's body with current—more than six amps—you'll cook the meat on his body. It's like meat on an overcooked chicken. If you grab the arm, the flesh will fall right off in your hand.... Presumably the state will return the remains to the person's family for burial. Returning someone who has been cooked would be in poor taste."[2]

Leutcher has redesigned electrocution systems in the hopes that they will be more efficient and cause less suffering. Unlike the commonly used pattern of eight brief bursts of electricity (in alternating five- and twenty-five-second increments), Leutcher's newer systems supply two one-minute blasts of current. The initial 2,640 volt blast creates in the prisoner's body an overwhelming rush of adrenaline. This adrenal storm usually keeps the prisoner's heart going after the first jolt. However, no human heart can survive the second 2,640 volt shock. To allow the adrenaline to dissipate, Leutcher's chair is programmed electronically to delay ten seconds before delivering the second blast.

Many medical experts believe that prisoners being executed, even under the most carefully controlled conditions, feel themselves suffocating—as the lungs refuse to function—and burning to death as body temperature rises above the boiling point of water. Harold Hillman, a British physiologist who has studied many executions, describes the effect this way: "It must feel very similar to the medieval trial by ordeal of being dropped into boiling oil."[3]

Willie Francis, a seventeen-year-old murderer, was sentenced to die in the electric chair in 1946. But because of a malfunction in the wiring, not enough current passed through his body to kill him the first time.

---

2   Susan Lehman, "A Matter of Engineering," *Atlantic Monthly*, Feb. 1990: 26–29.

3   Weisberg, "This Is Your Death."

He described the sensation later of feeling as though his brain were on fire and his lungs were frozen shut. "My mouth tasted like cold peanut butter. I felt a burning in my head and my left leg and I jumped against the straps."[4] A year later, after much legal wrangling, he was strapped into the same chair again and successfully executed.

Divorced from the emotional and ethical aspects of the matter, electrocution can be pictured as a purely physical process. The body—seen as a conductor of electricity—is a leathery bag containing a solution of electrolytes. Though electricity does not move in a perfectly straight line as it passes from entrance to exit, the greatest density of current is along the line connecting the two points of contact. But because the human body is a complex object for the current to pass through—unlike a uniform substance such as copper wire or salt water—the actual resistance of the body may vary greatly during the time the electricity is moving through it. The effects of the shock are often impossible to predict.

To make electrocution as efficient and expedient a process as possible, certain techniques of preparation have been developed. Like a patient being readied for surgery, the prisoner to be executed goes through an exacting process before the actual procedure occurs. Very important is the maximizing of contact. The prisoner's scalp is shaved down to stubble; a safety razor is used then to clear a spot at the center of the head. This is the place where the soaked sponge of the death cap will make contact. Similarly, an area approximately six inches

---

4   Weisberg, "This Is Your Death."

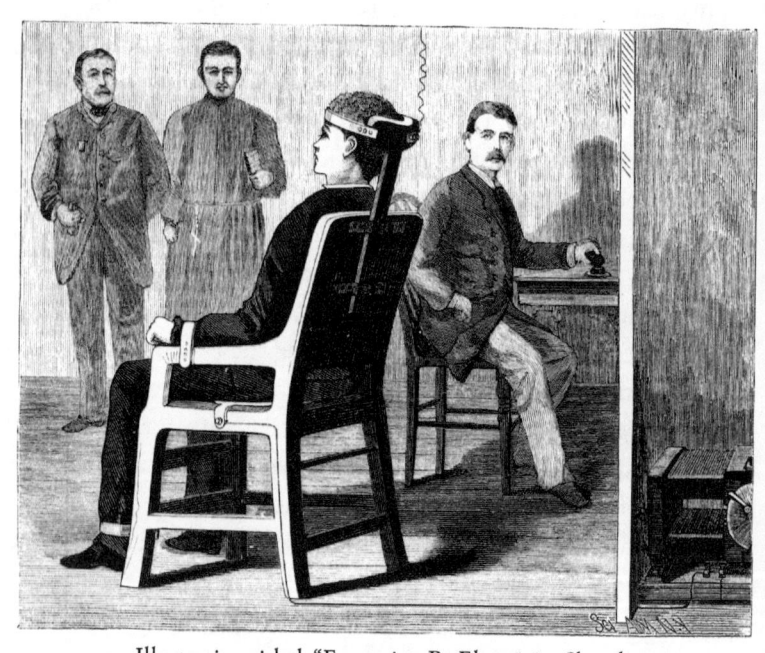

Illustration titled "Execution By Electricity Shortly To Be Introduced in NY State" from the June 30, 1888 issue of *Scientific American*.

above the ankle is shaved, to make the optimum connection with the ground pad.

Because prisoners usually lose control of their bladder and bowels during execution, they wear special diaper-like garments with tight elastic around the waist and leg openings and a thick cotton-fiber crotch.

Fred Leutcher believes that the mask the prisoner wears "allows the executee to enjoy some degree of privacy during the execution."[5] But it also enables the technicians and witnesses to keep an emotional distance from the process. Grotesque facial contortions, inhuman noises, drooling, vomiting, and the occasional occurrence of the eyeballs being squeezed out of their sockets—all of these can be safely hidden behind the leather mask.

A prisoner in the electric chair resembles nothing so much as a newborn baby: bald, gurgling incoherently, diapered, blind, and utterly helpless. But for the people who make the execution happen, the prisoner is even less than this. He's not a human being anymore, but a badly designed element in an electrical circuit. Fragile, with a wide diversity of tolerances and specifications, the human body is the least predictable component in the system.

Everything possible is done to ensure that the mechanism works as desired. The connection at head and leg soaked with conductive Electro-Creme—or paste-like brine solution—is the most efficient way of transferring electrical current into the body. Voltages and amperages are finely calibrated. The system itself is checked and

---

5   Lehman, "A Matter of Engineering."

rechecked, tested and inspected. Hundreds of previous executions give the prison personnel a good idea of what to expect. A controlled environment, witnesses, accurate analytic tools, the frequent presence of doctors and nurses lend the execution the air of a scientific experiment. But the body is always a variable.

Strapped into the chair, connected at head and foot, the body is—at least while the current runs through it—as much a part of the system as the lines that bring the power in from the generator or the insulation that surrounds the chair. However, the body is a makeshift component, an organic resistor whose rating in ohms can only be approximately determined. Though dissimilar in almost every other way, the body can be seen for those few moments as a grossly inefficient heating coil, like those in an oven or a toaster. The body in the chair is the most overt, the most extreme, example of the intersection of electrical power and human flesh.

The method of execution a society employs is a reflection of that society's vision of what it means to be human. The story of the first use of the electric chair sheds some light on our culture's attitude toward the always-uneasy relationship of body and mind. On August 6, 1890, William Kemmler, a murderer from Buffalo, New York, was the first person ever to be put to death by electricity. At this time, electricity was thought capable of creating life, transforming it, as well as destroying it. During this period, a scientific revolution was picking up momentum, and would change radically our culture's view of itself. The confluence of forces that brought the chair into existence—science, government, the legal profession, business, medicine—was unprecedented. And the results of Kemmler's death, for the popular imagination as well as the criminal justice system, continue to resonate to this day. How we see and define ourselves was influenced by, crystallized in, this most extreme conjunction of human flesh and electrical energy.

BUFFALO EXPRESS.

July 17, 1887

ELECTRIFIED CURS
27 OF THE UNMUZZLED EXECUTED ONLY A FEW ARISTOCRATS LEFT

The morning was an eventful one in the history of dogdom.

Twenty-seven luckless captives, whose term of probation had passed, were offered up on the electric altar. The new form of execution dispenses altogether with the "dull thud," the "sharp report," and the "loud splash." One by one the doomed dogs were led from the kennel room to the chamber of death. One by one they were placed in a box about 2 x 3, lined with tin, with about an inch of water in the bottom. One by one they were muzzled with a wire running through the mouth. A simple touch of the lever—a corpse.

The work of extermination was witnessed yesterday by Drs. McMichael, Wende, Park, Fell, and others, all of whom expressed delight at the expedition with which the work of destruction was performed. At present only three or four dogs of evident good social standing remain at the pound.

The fresh crop will probably be harvested tomorrow.

Chapter One

THE DEATH COMMISSION

The first instance of humans killing other creatures with electricity occurred in 1745. In that year—in two different locations: Leyden, Holland, and in Prussia—a device for storing static electricity was invented. The Leyden jar, as it came to be known, consisted in its earliest form of a glass vial partially filled with water, a cork in the bottle's mouth through which a nail or wire protruded into the water. By means of a friction device such as a rotating sphere of amber or sulfur to build up static electricity, the Leyden jar was filled with a charge. Though the device was crude and poorly understood at this time, a surprisingly large amount of energy could be stored this way, and released at will.

Abbé Nollet, who had earlier amazed the French court with his experiments using static electricity, was the first to demonstrate the Leyden jar in Paris. In public exhibitions he would discharge the jar into small animals, birds, and fish, killing them instantly. A craze soon swept the city, and inventors found new ways of using the jars as practical jokes as well as concealed weapons. A walking stick was created that could shock passersby, particularly useful on persistent beggars.

Besides being used as a weapon, Leyden jars also became popular in a kind of party game. People flocked to Nollet to be given the delicious jolt. And here already we see the odd mystique of excitement and danger that developed around electricity. The jars, and more sophisticated versions of the technology, soon entered the popular imagination as an intensifier for human experience. The novelty of the shock, the quasi-sexual mixture of pain and pleasure, the unknown nature of the discharge, all contributed to its allure.

An engraving from this period called "The Electric Kiss" shows an attractive young woman with her lips inches from her lover's while a second man—the scientist—cranks a static generator to fill her with an electrical charge. There's longing, excitement, intense anticipation, in the eye-to-eye connection between the two lovers. The erotic power in the scene is unmistakable, the lovers' passion palpable as their lips come close, closer, close enough for the spark to jump the gap.

Another illustration from the same period—"Cupid's Arrow Electrified"—depicts a similar situation; however, in this case there are three winged cherubs in the room to facilitate the sexual jolt. One cranks the generator and another holds a charged arrow while the lovers approach. Again, there is an electrical scientist in the scene, supervising. On the far right a third cupid looks on approvingly.

Electricity's erotic quality (erotic in both the narrow sense of sexual energy and the broader meaning having to do with natural forces of attraction, the essences of life and generation) can be seen in these early applications. As a love-toy, a weapon, and a force that could perhaps instill life in nonliving matter, electricity was conceived of as an elixir—mysterious, powerful, unpredictable.

When Luigi Galvani, in 1786, discovered that electricity applied to severed frogs' legs could make them kick, he opened a new world to science and the imagination. He believed that he had stumbled onto the supreme "life force," and in 1791 he published his "animal electricity" theory. During this time, when reason and science were believed capable of uncovering all of nature's secrets, it wasn't such a great leap of faith to accept his explanation. And to a certain degree Galvani was correct; electrical impulses are one of the basic constituents of organic life.

The idea that electricity could create life, or transfer life from one being to another, quickly took hold in the popular mind. Mary Shelley's *Frankenstein* is perhaps the best-known instance from this period. Cobbled together from decaying body parts, Frankenstein's creature is brought to life in a vaguely described process that was certainly influenced by Galvani's theories. In *Frankenstein*, the erotic qualities of electricity are muted, but longing and loneliness pervade the book. The preoccupation with generation, creation, and romantic love make an interesting complement to the grotesque science that Frankenstein employs. Rotting corpses come to life while virginal young lovers must die.

Though Galvinism is never mentioned in the book, there's little doubt that Shelley, an educated woman with a great interest in science, would have known of it. Galvani's theories came to prominence in Great Britain in 1802, when Professor Giovanni Aldini—Galvani's nephew and disciple—connected a voltaic pile battery to the severed head of a freshly killed ox. Before a gathering of nobility, the animal's nostrils swelled as though they were enraged, its ears pricked up, the eyes opened, and the tongue stuck out. The next year Aldini performed a more notorious experiment when he applied electrical current to the body of a hanged murderer, Thomas Forster. His body quaked and his left eye opened, his fist clenched and his facial muscles contorted in a hideous grimace; this, hours after he'd been cut down from the gallows. In 1804, Aldini continued to experiment with corpses, causing them to shake and thrash like ineptly mastered marionettes. One's fist clenched and pounded the table he lay on. Another rose briefly, as though to walk out of the room.

The popular press of the time of course made much of these discoveries. The craze reached such a pitch that in 1804 the Prussian government issued an edict forbidding the use of executed criminals in electrical experiments.

Not everyone at this time accepted Galvani's theory of animal electricity, but it is without a doubt that his influence was widely felt. Electricity became a source of both awe and amusement, terror and titillation. And though it would be decades before it could be said that electricity was truly harnessed or even understood, the connection between the enormous power of the thunderbolt and the fecund shock of the electric kiss was growing clearer.

In 1844, Robert Chambers's book *Vestiges of the Natural History of Creation* became a bestseller, going through four editions in the first year and many more after that. In *Vestiges*, Chambers claims that a chemist in Britain had succeeded in creating life by running electric current through solutions of potassium silicate and copper nitrate. It was later shown that the "life" observed was an airborne mite that had entered the solution because of the chemist's lack of basic experimental precautions. Whether he did or did not create life is not the question. The fact that many people—scientists among them—believed it possible is what's significant. Electricity, like most scientific discoveries in their infancy, had become something to project fantasies onto. It was seen as an agent of numinous transformation, the kiss of life as well as the kiss of death, an eruption of natural excess which could create and destroy as it pleased.

By the late nineteenth century, some folkloric notions about electricity had been dispelled; still, many scientists clung to the belief that currents and charges could be applied effectively in just about any situation. In the early 1880s experiments were conducted with the express purpose of determining what would be the result of passing large amounts of current through living creatures. In Buffalo, New York, soon to be dubbed the "Electric City of the Future," Alfred Southwick noted the death of a local man who'd come in contact with the brushes on a generator—one of the first deaths associated with commercial electricity. Dr. Southwick (a prominent dentist, and later member of the governor's "Death Commission") proceeded to experiment on animals, to determine how much current was needed to kill them.

Not long after, a more public spectacle was performed at the winter home of Barnum and Bailey's Circus, in Bridgeport, Connecticut. Before a gathering of scientists, technicians, and newsmen, a battery of forty-two cells was used to test the reaction of various animals as electricity passed through their bodies. The experiment began in late morning and went until nightfall.

The first animal experimented on was a baboon. The head keeper and three assistants fought with the animal—which managed to bite one of the men and tear the clothes off all four—to tie it down in the experimental apparatus. A sponge, connected to a wire, was forced between the baboon's long snapping teeth and a second was attached to its paw. At first only two of the cells were connected to the circuit, but even this low-level current was enough to make the baboon shriek and thrash against the ropes. The experimenters increased the current, connecting more cells to the circuit. The animal's fury rose to a peak as the twenty-eighth cell was connected: biting and snarling, it let loose agonizing wails. As more cells were attached, however, up to the fortieth, the baboon's reaction ebbed. It was most likely stunned by the current, or so overwhelmed by the pain that it was no longer able to respond. When it was released it immediately came back from its lethargic state and attacked the nearest keeper with renewed rage.

A tame seal was next experimented on. As soon as the current was turned on, the seal thrashed from side to side, knocking over tables and chairs and for a brief time the keepers were unable to approach the animal safely. But it was eventually subdued and more current applied. Frantic, the seal gnawed through the wire and freed itself.

A newspaper account describes the next group of experimental subjects: "The small monkeys behaved very much like little children. The moment they felt the current they screamed and seemed [to be] undergoing agony. When the wires were removed they appeared puzzled and in three cases took up the electrodes as if to study them."[1]

Next, dogs were attached to the cells and similar results were obtained, though one of the subjects, having had current run through the base of its brain, began acting oddly after it was disconnected. It showed rabies-like symptoms within a half-hour and the keepers were then "compelled" to kill it.

The final subjects were the elephants, which the newspaper account claims actually enjoyed having current pass through their bodies, though the rubbing together of their legs and the squeals they produced might be seen as evidence either of pain or pleasure.

Behavioristic experiments using electricity on animals (fifteen years before Pavlov began his more famous work) occurred at Harvard in 1887. To pacify a racehorse named Gray Eagle, Professor R.H. Harrison devised an

---

1   *New York Times*, 3 Feb. 1889.

apparatus with which he could shock a horse at will. A battery was arranged in a cart. The wires led to the horse's bit, so that the electricity would enter its mouth whenever it disobeyed an order to stop or get to its feet. After the current was applied a number of times, the horse was deemed "cured" of its willfulness and unpredictability.

Only a few months before, experiments were taking place in Buffalo that would lead indirectly to the creation of the electric chair. After his initial experiments helping the American Society for the Prevention of Cruelty to Animals (ASPCA) in the killing of unwanted dogs, Dr. George Fell began work in which life-giving and life-taking were both involved. Noted as the inventor of the Fell motor—an artificial resuscitation device that provided forced respiration—Dr. Fell was also the unofficial executioner's assistant when the electric chair was first used. In July of 1887, the same month he was executing dogs en masse at the Buffalo pound, he came to national prominence when he used the Fell motor—consisting of a bellows, piping, and breathing valve—to restore the life of Patrick Burns, who'd been given up for dead. To test the device further, Fell performed another set of experiments. After chloroforming a large dog, he cut open its trachea and inserted a breathing tube attached to the Fell motor. Next, the chest walls were removed so that the heart and lungs were exposed. Keeping the dog breathing with the foot pump, Fell placed the dog in a zinc-lined box and the electrified muzzle was attached to its mouth. The experimenters watched the function of the dog's vital organs as the current passed through its body. Regarding the heart, Fell said, "the instant the circuit was made it ceased its action and became a mere mass of quivering flesh."[2]

In 1878, electric shock was used as a substitute for floggings in the Ohio State Penitentiary, where prisoners were forced to sit naked in three inches of water while current ran through it. At the same time that electricity was being used to kill and punish, it was also touted as a cure-all. Galvanic belts, electrical corsets, magnetic healing, and dozens of electro-therapeutic devices, were patented and sold as cures for everything from toothache to cancer.

Even Thomas Edison, who will be discussed later as one of the prime movers behind the creation of the electric chair, added his contribution to the electro-medical field. His Inductorium—a Ruhmkorff induction coil—was guaranteed to cure rheumatism, gout, nervous diseases, and sciatica. He patented a medicine that consisted of morphine, chloroform, ether, chloral hydrate, alcohol, and spices and assured the users that it would cure them of headache, neuralgia, and other nervous diseases. Naming the medicine Polyform—an electrical engineering term—gave the medicine an air of scientific prestige.

The medical world had embraced electricity as a new tool—capable of great things. During this period, it was thought of as a pure and pristine force, powerful but at

---

[2] *Report*, "Commission To Investigate and Report The Most Humane and Practical Method Of Carrying into Effect The Sentence of Death in Capital Cases," 17 Jan. 1888, 78. (Hereafter, Death Commission Report).

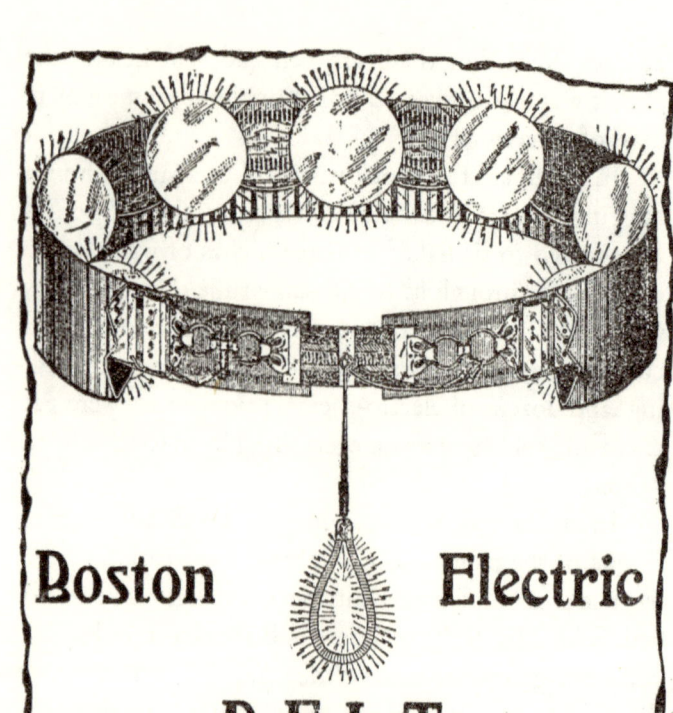

Invariably, there exists a person prepared to leverage the most recent fads to promote dubious medical practices.

the same time innocent. Its novelty certainly contributed to this perception; it had only been in the previous decade that electricity had truly become part of American life.

The number of medical people involved in the invention of the electric chair might seem surprising. However, at that time the alliance between medicine and the business of social control was more overt than it is now. And when the prevalent attitudes are considered (America in the 1880s was expanding rapidly, becoming a world power, taking its place as the preeminent source of technological innovation), it is easier to see why medical experts took a leading role the development of new forms of execution.

The concept of hygiene—the art of health—was pervasive in this period. Sanitary practices, the principles of right living to maintain one's health, were prominent in medical thought. Just as at the same time eugenic science was striving to purify the gene pool of America by preventing "dysgenic" procreation, cleansing the nation of "vicious protoplasm" by a program of hereditary hygiene, so the institution of capital punishment likewise was seen as necessary for social improvement. Few in positions of power seriously considered abolishing the death penalty. But with the invention of the electric chair it was increasingly thought of as a cleansing, not crude punishment. It became less a penal and more a medical procedure. Just as eugenics was thought by its practitioners to be humane and kind, so the use of electricity was thought to be a step forward, away from our violent, "uncivilized" past.

Brutal murders still occurred of course. If anything, those in positions of power saw the nation being steadily swamped in a tide of bestial crime. Their response, however, was not mere punitive retribution but what they considered a scientific, hygienic, humanistic answer. A dog was "cured" of rabies by electrocution the way a dysgenic person was "cured" of her "bad seed" by being sterilized.

The case of Roxalana Druse, from Herkimer County, New York, was instrumental in causing the medical and legal professionals to change their way of responding to violent crime. A popular culture heroine, a cause célèbre (her prison poetry published in many newspapers and her case reaching the highest levels of government), "Roxie" Druse was the figure who finally sparked the State of New York into adopting a new method of executing criminals.

Roxie Druse met her husband Bill in 1863; she was a hop-picker on his farm. Fifteen years younger than he, she pursued Bill with a tenacity that impressed all who knew the couple. She was of medium height, with a good figure and abundant long black hair. But it was her eyes to which the popular press attributed Bill Druse's downfall. "They were unusually large and black in color. They possessed a snaky glitter that was repulsive, while it [sic] also had a measure of attraction." They exercised over Bill "the fascination of a serpent."[3] She "caught" Bill Druse and asked him to marry her. At the wedding ceremony, he got cold feet and sat down, refusing to go on with it.

---

3   *New York Times*. 28 February 1887.

But one look from Roxie and he capitulated. She was a young woman noted for the rosy bloom of her cheek; however, after marriage the bloom quickly faded and she became notorious for the amount of makeup she used.

When her daughter, Mary, was sixteen years old, the house became well known in the area as a favorite haunt of male neighbors. Bill would go to bed early and the two Druse women would entertain late into the night. Bill complained of the laughter and sounds of drinking he heard, but Roxie and Mary rebuffed his every effort to curb their "orgies." Men slipped in and out of the house, a great deal of liquor was consumed, and Roxie's bad reputation spread wide. Eventually, Bill began to threaten reprisals.

A neighbor, Charles Gates, presented Mary with a nickel-plated, ivory-handled revolver, but at her mother's suggestion, she kept the gift hidden from her father. On the morning of December 18, 1884, another of their many rows began, Bill complaining that the fire in the stove was too hot. The family was eating breakfast when Roxie ordered the two boys—son George, aged 10, and the neighbor boy, Frank Gates, aged 14—out of the house. The boys were gone only a short time when they heard noises they later described as sounding like chairs being thrown.

As soon as George and Frank were out of the house, Mary had come up behind her father and tightened a length of rope around his neck. Roxie pulled the revolver out of her apron and fired two shots into her husband. He struggled, Mary pulled the rope tighter and Bill fell to the floor. Meanwhile, Roxie called the boys back in and placing the pistol in Frank Gates's hands, demanded that he fire the remaining rounds into Bill Druse. Frank backed away, shaking his head. "Do it, Frank. You must do it or you'll get the same." As Bill Druse lay on the floor quaking, Frank tightened his grip on the revolver and pulled the trigger three times.

Hearing her husband's groans and curses, Roxie bolted to the fireplace and took up the ax used for chopping wood. She hefted it, then brought it down on Bill's head. He begged her to stop. But she swung again, and again, and soon the head was cut clean from the body. Blood spurted fitfully onto the deeply gouged floor.

She saw there was no turning back now and she issued orders to her family like an officer commanding his soldiers. The boys were sent out for more firewood. Mary was ordered to cover all the windows with newspaper. Roxie wrapped her husband's head in her apron and placed it in the corner. When the fire in the old black-bellied stove was blazing, Roxie began the long and arduous work of cutting her husband into pieces and feeding him a little at a time into the flame. Using the ax, as well as a jackknife and Bill's razor, Roxie first cut his legs off at the knees, then the thighs. She severed his trunk into four sections and over the next twelve hours burned his remains to ash. The dog, meanwhile, had gotten into the room and was tracking blood all over the house. Roxie tied him up and when the body was finally dispatched, washed the floor and then spilled a bucket of paint on the makeshift butchering floor.

The ashes and the few bones that hadn't been burned were carried away from the house and buried in a clump of bushes. The wooden parts of the ax and jackknife were also burned and the metal blades sunk in a nearby pond.

Threatening the boys that they would meet a similar fate, she then told them, "If anyone asks where he is, you're to say he took a trip to New York City to visit his brother."

That night, Mary and Roxie entertained their gentlemen friends as usual, and when they were asked about Bill, they were vague and evasive in their answers. Suspicions quickly grew among Roxie's neighbors. And when one of her gentleman callers was accused of killing Bill Druse, his denials brought the issue to the fore.

Eventually Frank Gates did tell all that he'd seen on the morning of December 18 and lawmen came to arrest Roxie. Confronted by Frank Gates' story, Roxie threatened the fourteen-year-old with a suit of slander. But when he mentioned that she'd wrapped Bill's head up in newspaper, young George Druse said, "That ain't so, she wrapped it in her apron." This was enough, and Roxie and her daughter were taken away.

She was convicted and sentenced to die on the gallows. Appeals followed and her execution was postponed. By this time, news of her crime and the story of the victim had reached the wider populace, mostly through sensationalized newspaper accounts. Petitions began to flow into Governor David Hill's office, requesting that Roxie's death sentence be commuted to life imprisonment.

In the eleven-page order of respite the governor

issued, he explains why he postponed the execution and describes the variety of petitions that had come to him regarding the Druse case. First there were letters from those who were opposed to capital punishment in general and asked him to grant clemency. The second group asked him to commute her sentence based on her sex. There was strong feeling at this time that the death sentence should not be applied to women purely because it was unseemly, against nature, for women to be hanged. Some said there was precedent for this, claiming that Grover Cleveland (the former governor, and the only United States president to perform the duties of the hangman, while sheriff of Erie County) had commuted the sentence of a murderess, Mrs. Haight, simply because she was female. In Cleveland's written statement to the legislature, he admits that there is "naturally a feeling of repugnance against the execution of women,"[4] but then explains his commutation was based on the prisoner's advanced age and impaired health, not on her sex or the circumstances of her crime. The third class of petitions came from those who sought greater political rights for women and insisted that since women had no voice in making the laws that apply to murder and its punishment, then such laws should not be applied to women. Hill argues this away by comparing the position of women in New York to that of Indians who have no part in the creation of state law or "any political rights whatsoever."[5] He then draws on a moral argument: "The law

---

4   Public Papers of Governor David Hill, 1886, 324.
5   David Hill papers, 323.

which forbids murder, and declares that the life of the offender shall be forfeited, is Divine, as well as human law. The law of our State made by men is simply in harmony with the law of the Almighty."[6]

A fourth group of individuals wrote to the governor basing their claim on the facts of the Roxie Druse's life. Sympathetic to a woman who'd lived her life in apparent poverty, violence, and squalor, and bolstered by the claim that if she hadn't killed Bill Druse he would have killed her, they hoped the governor would show leniency. Governor Hill refused to alter the sentence and on February 28, 1887, Roxie Druse was hanged in the Herkimer County Jail, the first woman in thirty-nine years to be executed in New York State.

There was a military display that morning, the Remington Rifles of Mohawk marching the frozen streets of Herkimer. It was bitterly cold, near zero, and though the snow lay in deep drifts, still the town was crowded with visitors. Over a hundred men waited outside the county jail. Inside, Roxie Druse prepared herself for the execution. She presented a lock of her hair to one of her guards, and a poem to her spiritual adviser, Dr. Powell:

> Who will care when I am gone
> and the bird's music is hushed
> in the twilight dim and gloomy?
>
> Who my name will softly whisper
> who for me will kindly pray

---

6   David Hill papers, 323.

> when at last death has its way?
>
> Lying on my narrow bed
> who will smooth my dying pillow
> who will care when I am dead?[7]

She wrote a last letter to her daughter and ate the bowl of soup laced with opiates to make her more tractable prior to hanging.

The dress she wore to the scaffold was described in great detail by the newspapers, as though in a bizarre inversion of a wedding notice. The skirt was especially narrow, made for such occasions to maintain propriety as she swung at the end of the rope. The bodice too was tight and the cuffs on her sleeves were cinched with white ruffing—the only contrast to the stark black of the dress. At the front and at the top of her bodice she wore a bunch of roses, taken from the bouquet her daughter had sent. A small black cap and mask, with an elastic strap, hung down behind her head. Her hair was still long and deep raven black.

Twelve "reputable" citizens were there as witnesses, as well as seven deputy sheriffs and a large number of reporters. Roxie Druse wept softly as the death sentence was read aloud, then she was marched through the prison to the courtyard. Holding a bouquet, she was led up to the gallows. Dr. Powell, like a father there to give away the bride, accompanied her in the procession. The gibbet had been painted white the day before, and now

---

[7] *New York Times*, 1 Mar. 1887.

reflected the harsh February sunlight.

It was almost noon when Roxie was led to the gallows. She looked left and right, apparently oblivious to the intense cold. Roxie knelt and Dr. Powell gave a long and impassioned prayer, asking for leniency, and if that was not forthcoming, then at least forgiveness. Then the deputies stepped forward and pulled the black cap over Roxie's head. She moaned, cried, and then began shrieking, the sound muffled by the cloth that now covered her face. The noose was placed around her neck.

The trap fell at 11:48. The weight dropped and lifted her 3 feet off the ground. It took almost 15 minutes for her to die. She was not killed by a quick snapping of the neck, as was usually the case, but hung, conscious for a quarter hour, kicking and writhing at the end of the rope. The end came as she choked to death in her own spittle. She was declared dead at 12:03. The body hung on the gibbet until 12:14, almost a half-hour. After examination, the physicians stated that her neck was not broken and that she had died by strangulation.

This event was the impetus needed to get the New York State legislature to act. With the protest and public revulsion, with the legislators' sensibilities deeply offended by Roxie Druse's death, talk became much more serious about finding a new way of executing criminals. Dr. Southwick had already persuaded his friend State Senator Daniel McMillan to speak to the governor about a new method of capital punishment. But as with all changes in the technology of execution, the drive to find a new method was motivated more by consideration

An illustration from *An Innocent Woman Hanged: The Druse Case*, Philadelphia, Old Franklin Publishing House [1884?].

for those dispatching the prisoners than for the prisoners themselves. That Roxie Druse died slowly and painfully was a genuine concern for the legislators, but that she died shamefully, that her death reflected poorly on the State of New York, was a far greater problem.

Add to the image of a person both lionized and demonized, drowning in her own spittle at the end of a rope, the dogmatic faith in science and progress so prevalent in the 1800s and a very potent product is the result.

In *Violence and the Sacred*, René Girard argues that execution serves as human sacrifice, a method of redirecting the violence innate in human relations. "The purpose of the sacrifice is to restore harmony to the community, to reinforce the social fabric. The sacrifice serves to protect the entire community from its own violence; it prompts the entire community to choose victims outside itself. The elements of dissension scattered throughout the community are drawn to the person of the sacrificial victim and eliminated, at least temporarily, by the sacrifice... The purpose of the sacrifice is to restore harmony to the community, to reinforce the social fabric."[8]

Expanding on Girard's model, it can be argued that we as a culture, as a nation, kill those who represent the unacceptable possibility that we might commit the acts that put the offenders in prison. It's been suggested that we see in the criminal the unbearable desires and weakness that are so often found in ourselves and then in an act of mass catharsis, kill them to kill the uncomfortable emotions they evoke.

---

8   René Girard, *Violence and the Sacred*, 8.

All the changes in the methods of execution reflect changes in the way a society sees itself. Talk of humaneness in execution is an act of self-delusion to hide a deep discomfort. As will be seen shortly, various alternatives to the gallows were proposed. And all but the electric chair were rejected—supposedly because they were inhumane. More accurately though, they were unacceptable because they said something about late-nineteenth-century America that it did not want to hear. What New York (the Empire State, the most prosperous, populous, and powerful state in the union) wanted was a way to enhance its prestige. Its goal in doing away with the gallows was to further its image as being progressive, reformist, and at the forefront of cultural evolution.

The governor and legislature of New York needed a way in the 1880s of ridding themselves of the shame of brutal, obscene executions. As with the gas chamber in the 1920s and lethal injection in the 1980s, the electric chair was a way of redefining the state-as-executioner in its own eyes as well as the eyes of the world.

In 1885, in his first annual message to the legislature, Governor David Hill stated:

> The present mode of executing criminals by hanging has come to us from the Dark Ages, and it may well be questioned whether the science of the present day cannot provide a means for taking the life of such as are condemned to die in a less barbarous manner. I commend this suggestion to the consideration of the legislature.[9]

---

9   *New York Times*, 5 June 1888.

Not long after, the governor commissioned three people to "investigate and report at an early date the most humane method known to modern science of carrying into effect the sentence of death in capital cases."[10] The appointees were: Dr. Alfred Southwick of Buffalo, Matthew Hale of Albany, and "Commodore" Elbridge Gerry.

Southwick's role in the creation of the electric chair is seminal. Though a dentist, his work as a self-styled "penologist" gained for him the titles "Old Electricity" and "The Father of Electrocution." Like many others involved in the creation of the chair, he was thought of as a great humanitarian. Just as Dr. George Fell was involved in both electrocution experiments and the invention of life-saving devices (after the sinking of the *Lusitania* he developed an improved life-preserver) so Dr. Southwick believed he was working for human progress on a number of fronts.

He organized the dental department at the University of Buffalo, was well known as being one of the most skilled oral surgeons of his time, especially at the reconstruction of cleft palettes, and was president of the New York Dental Society from the late 1870s until 1895. Proud of his Quaker heritage, he was a dour, grim-faced man with a wiry, mustacheless beard and a grim, almost-Mosaic expression. He did in fact think of himself as a prophet, though the message he brought was not one with an overt religious content. Like almost all the major figures in the eugenics movement of the time, who came from Christian backgrounds but who threw off the church and substituted for it a religion of scientific

---

[10] Death Commission Report, 3.

perfectionism, so too Southwick pushed his vision of an improved human race with a religious zeal.

By the 1880s, Darwinian thought had begun to seriously erode the underpinnings of blind-faith Christianity, and technology had shown itself capable of doing what was thought until then impossible. So it's not surprising that many individuals were beginning to embrace what might be called scientific fundamentalism, substituting the tenets of science for those of religion, but retaining the moral and psychological vision that Christianity had promulgated for centuries.

The first to begin the push for electrocution, Southwick worked doggedly to influence public opinion as well as political leaders. Shortly before the Death Commission's report was made public, he was quoted as saying that the electric chair was "the end toward which I have been working for six years." And hoping that New York would be the first to use electrocution, he said, "I have noticed that the bill introduced in our legislature last year was copied in Paris... Germany has taken up the question, and I have just read that in New Jersey, attention has been called to our agitation of the matter. I wish that the Empire State would take the initiative in this step toward a broader humanity."[11]

The second member of the commission, Matthew Hale, seems to have been the least important.

Elbridge Gerry is perhaps the most curious of the three men to sit on the commission. The grandson of a designer of the Declaration of Independence, Gerry

---

11   *New York Times*, 24 Jan. 1887.

was what might be called a gentleman lawyer, virtually retired from the field by the time he was asked to chair the commission. His wealth and position were substantial; he was a member of many exclusive social organizations in Newport, Rhode Island, Martha's Vineyard, and Tuxedo, New York. With a nationally renowned collection of law books, an esteemed name and the title of Commodore of the New York Yacht Club, a position he held from 1886 to 1893 (Gerry's yacht, *The Electra*, was outfitted with every electrical device and appliance then available), Gerry gave an air of propriety, weight and dignity to an affair that might otherwise have been seen as mere titillation and sensationalism. But the aspect of his professional life that adds a peculiar twist to his appointment was his great interest in "humane" and protective efforts on behalf of animals and children. When the ASPCA was founded, he acted as legal counsel and was instrumental in the passage of many New York State laws governing the treatment of animals. In 1874 he was a strong proponent and one of the founding members of the New York State Society for the Prevention of Cruelty to Children (SPCC), whose branches were at the time often called Gerry Societies.

The laws of both protective organizations were shaped in large part by Gerry's work—he was a noted authority on canon and ecclesiastical law—and five years after its founding, he became the president of the SPCC. His influence was great enough in the legislature that he was able to procure laws conferring on the society corporate power to enforce the law.

The commission that he chaired—though they spent two years on research and discussion, and issued a ninety-eight-page report—was to a large extent merely a method of granting legitimacy to a foregone conclusion: that hanging must be abolished and electrocution used in its place.

The Death Commission began its work by sending to judges, district attorneys, sheriffs, and physicians a circular containing the following questions:

> FIRST. Do you consider the present mode of inflicting capital punishment objectionable?
>
> SECOND. Were you ever present at an execution, and if so, will you kindly state details of the occurrence bearing on the subject?
>
> THIRD. In your opinion, is there any method known to science which would carry into effect the death penalty in capital cases, in a more humane and practical manner than the present one of hanging?
>
> FOURTH. The following substitutes for hanging have been suggested to the commission. What is your view as to each?
> 1. electricity
> 2. prussic acid or other poison
> 3. the guillotine
> 4. the garrote

FIFTH. If a less painful method of execution than the present should be adopted, would any legal disposition of the body of the executed criminal be expedient, in your judgment, in order that the deterrent effect of capital punishment might not be lessened by the change?[12]

Roughly two hundred replies came back to the commission. Eighty of the respondents were against a change, eighty-seven favorite electrocution, eight poison, five the guillotine, four the garrote, and the remainder were noncommittal.

The Death Commission's report was transmitted to the legislature on January 17, 1888. A document more notable for its length and attention to grisly detail—especially the grotesqueness of executions in "barbaric nations"—than for intellectual rigor, it did have the effect of spurring the state to move more quickly toward electrocution.

The commission began by outlining the moral and religious background for capital punishment, enumerating all the instances in the Bible's Old Testament where execution is prescribed. After describing its use in other ancient cultures, the commissioners then skip to England and give a lengthy history of execution there. Again, the idea of progress and the advance of civilization is the primary concern.

The next section of the report is perhaps the most

---

12   Death Commission Report, 81. (All following references in Chapter One are from the same source.)

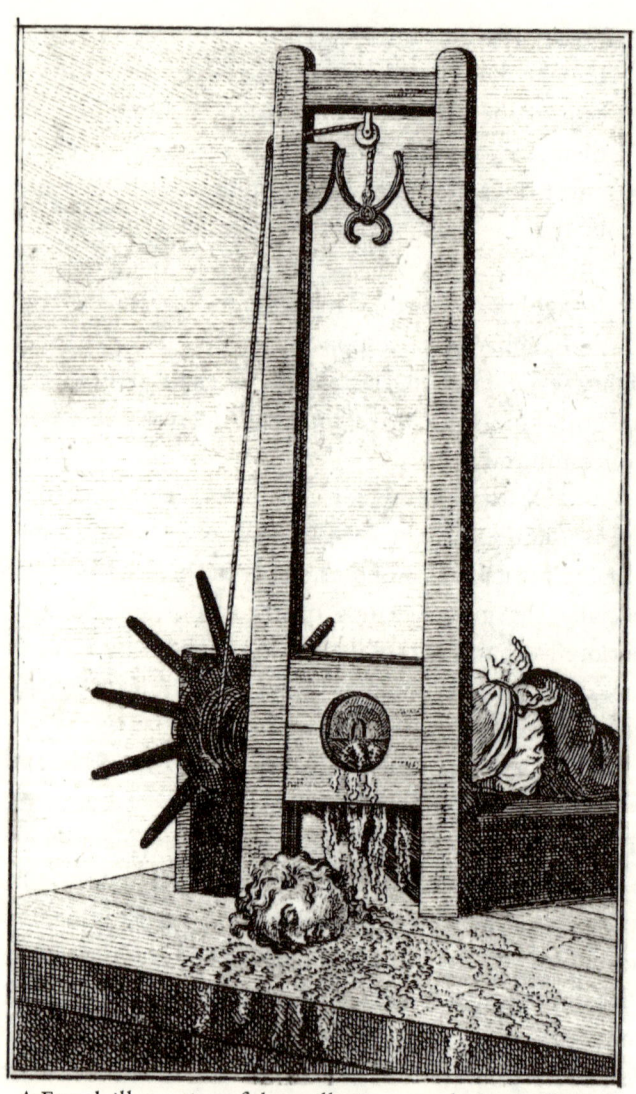

A French illustration of the guillotine at work, date unknown.

curious: a thirty-four-page world tour of execution. From the common—hanging, shooting, beheading—to the obscure—being blown out of a cannon, pushed off a cliff, disemboweled, and even transformed into an "illuminated body" by having holes bored in the flesh, filling them with oil and lighting them to make a human candelabrum—the subject is treated with obsessive devotion to detail. The main point seems to be showing most methods of execution as inhumane and savage, the province of "primitive barbarians" and religious fanatics.

Following this, the commissioners describe the "Present Methods of Execution in the Civilized Countries." From the long catalog of capital technologies, they decide that only five are used in the "non-savage" world: the guillotine, the garrote, the rifle, the scaffold, and the headsman's ax. One at a time, they analyze these methods and declare them unacceptable for the State of New York. The guillotine, it is admitted, is "apparently the most merciful, but certainly the most terrible to witness." The main objection they make to the use of the guillotine is that the "profuse effusion of blood which it involves." It is "needlessly shocking to the necessary witnesses and when the public are permitted to see the details the injury to humane feelings is unquestionable and unbalanced by any competing advantage." The commissioners assert that a method of execution that involves "the fatal chop, the raw neck, the spouting blood," is "very shocking to the feelings, and demoralizing, as such exhibitions cannot fail to generate a love of bloodshed among those who witness them." Though death is, they concede, instantaneous, painless, and beyond all possibility of resuscitation, the guillotine is still not a method acceptable because it is associated with "the bloody scenes" of the French Revolution, and would be "found totally repugnant to American ideas."

The second method considered is the garrote, an iron collar tightened around the neck by a screw mechanism. This method is speedy, when performed by an expert executioner, but the commissioners reject it because of its associations with Spain, the only nation than using the garrote. It was thought to smack of medievalism, the Inquisition, torture, and "un-American ways."

Shooting is treated next. The objections the commission makes to the rifle are ill-defined. Though neither cruel (causing instantaneous death) nor unusual (being used frequently in military settings), shooting is rejected because of its associations with "military despotism" and because it would be "demoralizing, particularly because of its tendency to encourage the untaught populace to think lightly of the fatal use of fire-arms."

The headsman's ax is barely mentioned. Apparently the objections to the guillotine were thought applicable to the chopping block as well.

Next, the report devotes twenty-three pages to the subject of hanging—in particular, the ways in which it can be botched. The question of liquor and hanging is brought up: for the condemned prisoner as well as the executioner, and the commission uses this to paint the institution of hanging as crude, and somewhat clownish. "Drunkenness is in itself an offense," the report declares.

"In the moral aspect the gross impropriety of sending a man into the presence of his maker intoxicated is too obvious to require comment."

Attempts made by the condemned to commit suicide are described at length. On the rare occasion when the prisoner is able to slit his or her throat and still lives, there's often the added embarrassment and awkwardness of the prisoner's head coming free from the body, or even, as is described in vivid detail, if he dangles from the rope and is still able to breathe because of the cut below the level of the tightening noose. Case after case of bungled hangings are related, and the objections to hanging are then enumerated:

> First, resistance or suffering by the offender; second, unskillfulness or brutal indifference of the executioner; third, misconduct of bystanders; fourth, sympathy of bystanders occasioned by the great age or other personal circumstances of the condemned; fifth, complication of the process by supposed necessity of executing more than one person at a time.

The report spends some effort analyzing the question of gender and execution, and why there is such a "grave, deep-seated objection which exists in the public mind against inflicting the death penalty upon a woman." Instances are described of a young girl struggling for ten minutes at the end of the rope, of a woman's clothes being torn from her body, of women becoming pregnant in prison, so that their execution would be postponed.

"It is not," the report claims, "so much to the execution of women as to the hanging of women that the general objection is addressed." The commission believed that the public's objection was a response to the "barbarity, the increased inhumanity, the heightened shock to the feelings, when the usual incidents of hanging as the mode are imagined as applied to a female." In other words, the emotions stirred—revulsion, titillation, excitement—reflected badly on a society that thought of itself as rational and on the upward path of progress. The commissioners recount scenes outside of prisons on the days when executions were taking place. Drunkenness was common, as were obscene language and "young country lads and girls clinging in reeling groups together, or even rolling in the gutters of the public streets." And when these "orgies" or "riots" were associated with a notorious woman doing the "scaffold dance," the bounds of propriety were too severely broached: execution as aphrodisiac, destroyer of public order and decency.

The commission clearly had its agenda set before beginning its research. Besides trying to find a method that would uphold New York's reputation for progressiveness, they also hoped to separate the United States from its medieval, autocratic, cousin-states in Europe. A method of execution was needed that was uniquely American and that drew on America's strengths: technological prowess, innovation, willingness to discard old ways.

It is curious that the commission would treat the subject of using "galvanic current" to resuscitate the executed prisoner and give some instances of when this had supposedly occurred, and then only a few pages later make the proposal that the rope should be replaced by electricity. Even for these men—convinced that the proposal would create the most humane and expeditious method of execution—electricity remained a mysterious force, capable of giving life to the dead as well as taking it from the living.

The report recommends without hesitation that "the time has come when a radical change should be effected." It quotes Elihu Thompson (an associate of Thomas Edison) as being strongly in favor of adopting the electric chair and that alternating, rather than direct, current should be used in executions.

After a brief discussion of the speed of electricity versus the speed of the impulses through the human nervous system, the report concludes:

> FIRST. That death produced by a sufficiently powerful electrical current is the most rapid and humane produced by any agent at our command.
>
> SECOND. That resuscitation, after the passage of such a current through the body and functional centers of the brain is impossible.
>
> THIRD. That the apparatus to be used should be arranged to permit the current to pass through the centers of function and intelligence in the brain.

The report then reprints excerpts from their

questionnaire responses, statements that seem more like satisfied-customer testimonials than scientific or legal arguments. Judge Vann, for example, says: "Electricity is not only the least painful and repulsive, but also the quickest, most certain and most easily administered."

Before the commissioners end their report with a model text they hoped New York would adopt as law, they devote a surprising amount of space to the question of the prisoner's body after execution. They evidence a great deal of concern that capital offenders are frequently treated as heroes after death, and that returning their bodies to family or friends allows the celebration of the criminal as an antisocial martyr. "Pageant obsequies permitted after execution often constitute a quasi apotheosis of the crime and invite the imitative ambition of less distinguished evil-doers." They then suggest that a "statute be enacted giving the bodies of executed criminals to medical colleges for dissection," and forbidding the description of the execution in the press. Their reasoning was that criminals "are often more concerned as to what will happen to the body after death than as to their future spiritual existence." Regarding the deterrent effects of capital punishment, they state that most criminals "would hesitate long to commit crime involving its application if they were certain that after execution their bodies were to be cut up in the interest of medical science." The dismemberment was then to be followed by a speedy burial, covering the body with a "sufficient quantity of quicklime to consume the body without delay."

This was the only element found questionable when

the report was submitted to the legislature on January 17, 1888. A rancorous debate ensued, Catholics arguing that the body should not be dissected, but buried whole in consecrated ground. Claiming that it's the state, not the family of the condemned, who owns the body, Protestant lawmakers succeeded in getting the bill passed with no substantive changes from what the commission recommended.

On June 4, 1888, Governor David Hill signed the bill into law, to take effect January 1, 1889, and applying only to crimes committed after that date.

Buffalo Evening News.

March 29, 1889

PROBABLY MURDER
A WOMAN BRAINED WITH AN AX BY HER HUSBAND THIS MORNING.
"I'VE KILLED MY WIFE."

A woman living at 526 South Division Street was hit on the head with an ax this morning and the surgeons at Fitch Hospital think she will die. Her husband inflicted the blows and has been arrested.

The man gives his name as William Kemmler, though it is doubted if that is right.

Five months ago, William Hort, 29 years old, his wife Tillie, 24 years old, and their daughter Ella [sic] four years old, moved into the rear rooms in Mrs. Reid's house

The couple did not live amicably together and Mrs. Reid frequently heard them fighting. Time after time blows were struck and Mrs. Reid frequently thought that one or other would be killed.

CHAPTER TWO

THE ARC OF THE AX

After work on Thursday, William Kemmler—also known as Billy Hort—and John "Yellow" DiBella, went salooning. They'd been at the beer off and on all day, but after quitting time, moved on to more serious drinking.

Billy lived only a short walk from the heart of Buffalo's red-light district, the Canal Street-Dante Place neighborhood. Within a twenty block area there were more than 75 brothels, 120 saloons and 19 free-theater saloons. Five cents bought a draft beer or a glass of home-brewed whiskey. Mixed drinks went from a dime to fifteen cents. Bottled beer cost a quarter. Music was provided by a slovenly trio in one corner, or on the kerosene-lamp-lit stage (out of tune piano, beer-sodden violin, and a cornet built for Civil War marching bands). In the basement places there was rat-baiting, with bets on which dog could catch and kill the most rodents in a three-minute round. Billy and Yellow avoided the German-style beer gardens a few blocks away. The music was slightly better (a small orchestra, singers sailing through melancholy and sentimental favorites) but the liquor was more expensive. They stuck to the places thronged with canal men and Great Lakes sailors.

On the cramped stages of the Lion, the Star, the Only Theater, Bonney's, or the Alhambra, they might see any number of entertainments. Comedy in the burlesque style one night, wrestling and boxing exhibitions the next, female dancers, and Billy's favorite: "gouging matches" in which stuporous combatants wore sharpened metal thimbles and tried to tear each other's eyes out.

Since coming to Buffalo eighteen months before, Kemmler had built up a moderately prosperous business selling vegetables on the street. He had half a dozen employees—huckster boys and wagon drivers—and though most of his cash was flushed away in bars, he'd managed to save up a few hundred dollars.

Billy and Yellow went from saloon to beer room, from brewhouse to wine shop, making small talk with friends and drinking cohorts. The liquor-crawl had a certain uneasiness to it. Billy might have been springing for all the drinks, but his tone was getting ugly, accusations and not-so-veiled threats seeping into what passed for conversation.

Yellow was one of Billy's wagon drivers, a younger man and better looking. The talk kept coming around to Billy's "wife"—Tillie Ziegler. Something was very much wrong and as the whiskey lubricated his tongue, Billy got closer to laying out the truth.

Tillie had left her husband in Philadelphia to run off with Billy. Fred Ziegler was a drunk too. Tillie met with Billy secretly, and taking her four-year-old daughter in tow, had fled with him four hundred miles north and west to Buffalo.

Now however, Tillie was getting increasingly impossible to live with. The arguments could be heard by neighbors and the landlady, Mrs. Reid. Tillie had made hints about leaving, going back, perhaps, to her husband.

Giving up for the night, Billy and Yellow stumbled from the saloon. The noise and stench of Buffalo's worst slum crowded in on them. Piles of manure from the stables stood so high they came up to the second-story windows. Smokestacks belched black and caustic fumes day and night. Far out-stripping in population Toronto, Niagara Falls, and Rochester, Buffalo was already the major industrial city in the region. Though by 1890 the railroad had made the Erie Canal obsolete, Buffalo's position—western terminus of the canal and eastern terminus for most Great Lakes ore and grain traffic—had encouraged rapid industrial growth. Over a hundred iron and steel factories were functioning already in Buffalo: blast furnaces, smelters, foundries and manufacturers of finished metal products. By the end of the 1880s, Buffalo was the primary challenger for Pittsburgh's role as the country's iron and steel capital. And by this time too, it was second only to Chicago as the United States' main railroad nexus.

Buffalo might not have had the hard-working reputation of Chicago or the allure of vice and violence of New York, but with its ramshackle slums and iron plants that lined the banks of the lake, it was definitely not a gentle or genteel place to live.

Billy Hort (as he called himself, taking the name of an acquaintance from Philadelphia) had fit in, flourished, in this world of whorehouses, saloons, dice shops, over-crowded tenements and collapsing wood-frame houses. He worked when he needed to, and had over the previous few months been drinking himself into a state of stuporous rage. Tillie's infidelities—whether real or imagined—had pushed Billy close to the edge.

As they made for home, conversation ran again toward women, Billy telling his younger companion over and over, "Women are hell."[1] As his memory blurred and his fears loomed larger, Billy began telling Yellow of the marriage he'd fled in Philadelphia. He'd been courting a young woman named Ida Porter. One night, she got him brain-dead drunk and persuaded him to marry her. The following day he deserted her. Two weeks later he'd sold his business for twelve hundred dollars and without speaking to any of his friends or family, had disappeared.

"Women are hell, Yellow. Women are just hell."

His threats and sodden ponderings had taken on a more personal tone by the time they'd returned to his place, Billy ranting about his wife and how little he trusted her with other men. "If she ran off with me, why shouldn't she run off with someone better?"

DiBella too was drunk, but not so far gone that he didn't notice the looks Billy was directing his way. Later, at the inquest, he admitted that that night he pushed his

---

[1] Dialogue in Chapter 2 is reconstructed from sources. See *Buffalo Evening News*, 29 Mar.-2 Apr. 1889.

bed up against the door, worried Billy might come in while he slept to take his revenge.

Billy woke early the next morning in the four-room flat he shared with Tillie and little Emma. Situated at the rear of the building, the flat was drab, dingy, and ill kept. Little light came in through the windows and the stove provided the only heat.

Hung-over, hungry and still haunted by the doubts he'd carried now for weeks, Billy stalked around the flat. He drank some lukewarm beer, gnawed a crust of bread, and looked out in the yard where one of his huckster boys, Dennis MacMahon, was already loading vegetables from the delivery wagons.

Emma was up now too, tottering around in a ragged flannel nightshirt. At four years old she already looked much like her mother: soft-featured, round-bodied, with a full head of dark hair.

Billy noticed that Tillie's travel trunk wasn't where she normally kept it. A big, leather-strapped case, it was all she'd taken with her when she fled her husband a year and a half before.

He tried the lock. It was unfastened. He hefted the lid and looked inside. All Tillie's belongings were packed away there: clothes, a missal, her hair brushes, a pair of shoes, a few papers, which Billy, being illiterate, couldn't read. He sifted through her things, feeling the rage returning, the black venom boiling up inside him. Hearing Tillie in the next room, he shut the trunk lid.

There was a knock on the door and Billy heard Dennis talking with Tillie. "Good morning, Mrs. Hort."

"Go down to the grocery for me, would you, Dennis? There's a few things I need you to buy."

Billy went into the main room and saw Tillie there with the boy. She gave Billy her oh-God-you're-drunk-again look and went on with her instructions to Dennis. The boy said good morning to his employer and, getting no reply, ran off.

Seeing Billy in his rage again, Emma scooted out of the way when he came at her mother. "I seen inside your trunk," he said.

Tillie didn't answer.

"It's all packed."

"I was just making it nicer. Straightening up."

"That's a lie. You're getting set to leave me."

"Don't be foolish. Why would I—"

"You're making to leave me. I know it!" He grabbed Tillie by the collar of her dress and pushed her up against the wall. "You're going to leave me like you left Fred."

"Billy! Stop it!" Tillie tried to get away but he pressed in on her, poking at her chest with his finger.

Emma stood in the doorway staring. She'd seen her "Papa" act this way before. She'd heard her mother and Billy shouting and carrying on too many times to count. He had Tillie now by the throat, shaking her and shouting, "It's Yellow, ain't it? You're making to run off with Yellow like you run off with me."

"Billy, don't be crazy. I'm not—" He slammed her against the wall, shouting back into her face.

Again the old recriminations came out: he was a stinking drunk and she was a whore, making eyes at any man who passed her by. Emma watched and waited for the two of them to calm down again. Billy would go for the bottle and her mother would sit awhile weeping, then get back to her work.

"Why is your trunk packed like that?"

"Billy, stop it!"

"I saw you and Yellow talking. I saw you smiling at him."

"If you don't stop right now I'm going to—"

"You took money from my roll. I had seventy-five dollars there. You stole my money."

"Billy, I did not!"

"Tell me the truth. You were making to leave me."

"If that's what you want to hear," Tillie shouted back, "if you really want me to say it then here it is. Yes, I'm leaving you this time. There! I said it. I'm leaving you because you're a no-good drunk. I'm taking Emma and getting her away from you. There! Is that what you want to hear?"

Billy let go of her suddenly. He turned and without saying another word went outside. Yellow and one of the other drivers were standing in the morning sunshine, waiting for Billy. He stalked past them to the barn where he kept the wagons.

In the cooler morning darkness of the barn, Billy felt the rage ebbing slightly. Pigeons fluttered overhead. The smell of hay and manure and heaps of feed were like a sedative—for a moment. They'd been through this all before, the arguments and threats. Was this different than any other morning?

Two renderings of the likeness of William Kemmler
from contemporary newspapers.

Then he heard Yellow talking with the other man, and Billy started to slip again. He stood a while in the shadows, peering out at the sunlit courtyard. The horse was shaking her head, twitching her tail halfheartedly. Yellow gnawed a plug of tobacco.

Billy went to the corner where he kept his tools. A shovel for mucking the stalls, a coal shuttle, a blacksmith's tongs. And a carpenter's hatchet with a wooden handle, stained by the sweat of his hands. The head was pitted iron, flat on one end to use as a hammer. The blade was wide like an Iroquois war ax. There was a notch in the underside for pulling nails.

He picked up the hatchet, hefted it to test its weight. He swung it and buried it deep in an oaken beam. He yanked it free, heard Yellow's laugh, his idiot mocking laugh, and went back outside.

As he went past, Yellow said something. Billy heard nothing but his own heart throbbing at the back of his neck. He kicked open the door and went inside the flat. The first room was empty. And the second too. Had she fled already?

No, he heard Tillie's voice from the kitchen. She was with Emma. Nonsense, small talk, trying to cheer herself up by cheering up the little girl. "Everything's going to be fine." Billy went across the room. Emma turned to look, saw the hatchet in his hand. But Tillie pretended she didn't hear him there.

"Mama," the little girl said. "Mama, you better—"

Billy lifted the hatchet.

"Mama." Tugging on Tillie's skirt.

Billy swung, feeling himself extended out through the ax shaft, the iron head, the glinting silver edge.

"Mama!"

The first blow knocked Tillie against the wall. Billy pulled the hatchet back and let fly again. Another and another, chopping ragged gashes in her scalp. Tillie fell to the floor, turned finally and looked up at Billy. The hatchet came down again.

Screams now, squealing like a pig being slaughtered. She held her hands up to protect herself, to beg him to stop, and again the hatchet came down, tearing back a flap of her scalp.

Billy had never felt so powerful, so alive. His whole being was captive in the arc of the ax, the glint of the blade. He'd given up his will, letting the ax, the screams, and the blood control him. The roles were reversed and he was now just an extension of the ax—crude muscle to propel it through the air, to make it do what it must.

Up and down the hatchet went, machine-like, relentless. Tillie crawled around the blood-slicked floor, moaning. Billy grunted with every blow—hard, heavy work.

In the corner Emma stood watching, mesmerized by the sight. Now gore was flying through the air, splashing Billy's shirt and pants, his hands and face. His eyes were bright, molten glass. He watched himself raise and lower the hatchet, he heard Tillie's whimpering, but it was as though he were trapped inside a machine running out of control.

All the arguments were done now, all the recriminations and beer-drenched threats. Now it was just the mechanical rise and fall of the hatchet. Twenty-six blows to the head. More to the arms, shoulders, breast. Tillie's hair hung down, blood-soaked. Her body swayed back and forth.

At the inquest, Mrs. Reid stated that it sounded like firewood being cut. She'd heard the argument from next-door, then silence, then the steady chopping sound.

He left Tillie there, still on her hands and knees, blood-smeared, mewling. He dropped the hatchet on the floor and went to Mrs. Reid's.

He pushed open the door. "I've killed her," he said, his voice dull and nerveless. Daubed with blood and brains, torn skin and hanks of hair, Billy stood swaying in the morning light.

"My God," Mrs. Reid said, "what did you—"

"I killed her. I had to do it. I meant to."

"What do you mean?"

"I killed her and I'll take the rope for it."

Billy went back to the apartment and Mrs. Reid fled to her neighbor, Asa King, for help.

Billy found Emma still standing in the room, paralyzed by what she'd seen. He picked her up and placed her in her little chair. He stood there a while, staring at Tillie's motionless body. Then, hearing voices, he pushed himself out of his stupor.

Outside, he was met by Asa King, who demanded that Billy go with him to the police station. Billy just shook his head and kept moving down the street.

He wandered a short while and ended up in Thomas Malone's saloon at the corner of South Division and

Jefferson. He went to the bar and called for a drink. King had followed him there and told the barkeep not to give Billy any liquor. Crusted with blood, dead-eyed and speaking in a whisper, Billy stood his ground. "Whiskey. Now."

The barkeep stared back at him. "What in the name of God—"

"Give me whiskey!"

At this point Officer O'Neil of the third precinct arrived at the saloon and told Billy he'd have to come along with him. Billy shrugged, giving up with no resistance.

At the station, he was questioned by Lieutenant Barry, but he'd say little now. He gave his correct name—Kemmler—and beyond this would say only, "I killed her with a hatchet. I wanted to kill her and the sooner I hang for it the better."

The first man to enter the flat at Division Street after Billy left was Dr. Blackman. The kitchen was chaos: blood and wreckage, scattered and broken dishes, overturned chairs. Lying near the stove was the motionless body of Tillie Ziegler, soaked scarlet from head to foot. Around her had formed a congealing pool. The walls and even the ceiling were spattered with her blood.

Dr. Blackman examined her and found that the heart was still beating. He tried to attend to the wounds on her head, but she was well beyond hope. He placed her on a bed until she could be moved to Fitch Accident Hospital.

Twenty-six cuts were found in her skull, ranging from one to four inches across. Surgeons led by a Dr. Park labored long to keep her alive, shaving her scalp and removing seventeen pieces of bone and a mass of brain tissue. She didn't recover consciousness after the assault and died near midnight.

When questioned by the police, Emma said only, "Papa hit Mama with a hatchet. He struck her in the head and she laid down on the floor."

The inquest was held three days later: Tuesday, April 2.

Various of the men who worked with Billy testified, as did the officers who'd been at the scene of the killing and dealt with him after he'd been arrested. The coroner made his statement. It didn't take long for the verdict to come back from the jury.

It found that "Matilda Ziegler came to her death at Fitch Hospital in the city of Buffalo, March 30, 1889 at 12:50 from wounds inflicted on her head with a hatchet in the hands of one William Kemmler, said wounds producing a comminuted fracture of the skull and being the immediate cause of her death.

"We the jurors from the evidence adduced that the said wounds were inflicted by the said William Kemmler with a deliberate and premeditated design of causing the death of the said Matilda Ziegler."

Kemmler was immediately arraigned in police court. He entered slack-faced and silent, handcuffed to Sgt. Dugan. The judge read the charge and asked how he would plead.

"Guilty," he said, barely audible.

As though still drunk and unsure where he was and what was happening to him, he allowed himself to be led back to the jail.

Buffalo Evening News.

April 4, 1889

ELECTRIC DEATH

THE HORRIBLE EXPERIMENT MURDERER KEMMLER WILL HAVE TO SUFFER IF CONVICTED. DISTRICT ATTORNEY QUINBY WILL MAKE A QUICK JOB OF THE SOUTH DIVISION STREET HATCHET FIEND.

"If Kemmler is convicted of murder in the first degree will he be killed by electricity and who will send the fatal shock?" Police electrician Plumb was asked yesterday afternoon by a News reporter. The indefatigable electrician hesitated a moment, then pulling his beard he remarked: "I think there will be some hitch in executing these brutal murderers by electricity. Nobody in the world has got enough power over electricity to tell what it will or can do. It's worse than a mad dog. Sometimes it will do one thing and the next time you don't know what it will do. These so-called experts say they know that they can kill a man instantly. I would like to know how they tell that. No man has been killed yet and how do they know it is going to act on the first victim? Sometimes you will notice that a man or woman is killed by electricity by accident. Then again they will only be shocked. How do you account for that? The accidents have been almost similar at times. They will put a murderer in a metallic chair and when the circuit is turned on it may only paralyze him. What a horrible death that would be. Electricity will paralyze and still not kill a man. So what will they do if the shock does not kill? Everybody remembers the way the Humane Society tried to kill dogs by electricity a few years ago. After some dogs received the shock, they kicked for fifteen minutes. Where did the humanity come in? The same difficulty may be encountered, and probably will, when the trials are made with the murderers. You can safely bet that there will be a hitch somewhere."

Chapter Three
THE WIZARD

"And Edison said, 'Let there be light!'" Millions of Americans heard these words via radio at the climax of the "Golden Jubilee of Light." The year was 1929 and by this time, only a few years from his death, Thomas Edison had achieved an almost god-like stature in the popular mind.

The jubilee took place at a reconstruction of Edison's Menlo Park laboratory. Henry Ford, one of Edison's most loyal disciples, had devoted millions of dollars and years of work to his Greenfield Village, a romanticized version of America as it "was and should be." And as one of the central attractions he had built a simulation of Edison's most famous research facility, which had burned to the ground years before. Ford collected scraps of wood, papers, discarded tools and equipment, even the red New Jersey clay from around the Menlo Park "Temple of Science" in order that his re-creation would have the same near-numinous atmosphere that the original building had. On October 21, 1929, at the heart of

Ford's grotesque vision of nineteenth-century America, Edison took part in the last great spectacle of his life.

The jubilee was conceived by General Electric as a way of deflecting negative publicity away from itself, for it was being investigated by Congress for monopolistic practices. Moved from the G. E. facility in Schenectady to Michigan, it attracted dozens of political and scientific notables. Even President Hoover was on hand to witness Edison reenacting his "creation" of the first incandescent bulb.

Edison was old and ill; he collapsed later at a banquet in his honor. But the ritual, the myth-making publicity and the sheer spectacle of the event were enough to coax Edison out of semi-retirement to participate one last time in an outlandish act of self-promotion. At one point, taking a trainboy's basket of wares, he clownishly tried to hawk them as he had sixty years before, shouting at the President, "Candy, apples, sandwiches, newspapers!"[1]

Graham McNamee, a well-known radio announcer, did the play-by-play at the peak of the festivities. Over 140 NBC affiliates carried the event live, the greatest radio hookup ever attempted up to that time.

"The lamp is now ready, as it was half a century ago. Will it light? Will it burn? Edison touches the wire.

"Ladies and Gentlemen—it lights! Light's golden jubilee has come to a triumphant climax."

And Thomas Edison's apotheosis as America's foremost inventor, thinker, and man of science was complete.

It is fitting that this event was pure simulation, an advertising ploy. Ford's kitsch replica village made a

---

[1] Robert Conot, *A Streak of Luck*, 444.

perfect backdrop for this showman's last act of re-creation. Edison was without a doubt one of the most gifted inventors of his century. However, the great amount of self-promotion and self-delusion that made their way into his biographies has blurred the line between mythic and historical truth. Edison claimed to have no interest in "Barnumizing," yet his greatest invention may have been himself.

He gave the media what they wanted: sensational stories, hot copy, hype, and hope. And they in turn gave him the opportunity to revitalize the mythos year after year. Edison the actual person and Edison the mass media creation are impossible to separate. A self-made man with little education, maniacally hard-working, driven not only to succeed but to defeat his opponents at every turn, a man who needed to be seen as the paragon of inventive genius as his abilities were eclipsed by others, Edison found himself trapped in a tangle of falsehood. He came to believe the public image, and its unreality—especially as he grew older—created constant frustration and frequent failure.

Edison stands as the one man who might be said to have invented the twentieth century. Motion pictures, recorded sound, the telephone, the distribution of electric power and its corollaries (commercial lighting, household appliances, home entertainment), radio, and public transportation were all shaped by Thomas A. Edison's work. He ushered in one of the twentieth century's most important dynamics—the primacy of technological image over natural reality—and at the same time was the chief prophet and icon of its underlying philosophy: the gospel of endless progress.

His most successful work was about extension: sending power through miles of wire, telecommunication across the nation. Photographs and motion pictures are a way of encapsulating sight and sound, preserving them, and transporting them elsewhere. Though compared to late-twentieth-century science much of his work seems infinitely crude, it was a watershed for Western technology. His career was a decades-long struggle to stretch the human reach through the use of electrical power. While he did expend some effort on antiquated technologies—his poured concrete houses and monstrous ore-crushing plant for instance—most of the work he's remembered for, and which truly changed the world, was at the micro level. Journalists intuited this early on, often remarking on how delicate, tiny, and sensitive his apparatus were.

He and his inventions were thought capable of wonders far beyond most people's imaginations. A *New York Times* headline has Edison proclaim: "Everything, anything, is possible." As Wyn Wachhorst points out in his *Thomas Alva Edison: An American Myth*, one of the most profound of American beliefs is that human effort and ingenuity, human will and desire, cannot be constrained. "Historically, Americans have tended to deny the reality and necessity of limits, a trait that has led to an apocalyptic view of conflict and refusal to acknowledge its natural role in human experience."[2] And Edison, as the one man who was thought capable of anything, did much to

---

2   Wyn Wachhorst, *Thomas Alva Edison: An American Myth*, 105.

crystallize and exemplify this belief, this crypto-religion of progress and constant improvement for humanity.

On April 1, 1878, the *New York Daily Graphic* printed this headline: "Edison Invents a Machine That Will Feed the Human Race—Manufacturing Biscuits, Meat, Vegetables, and Wine Out of Air, Water and Common Earth." Though it was an April Fools' joke, many other papers picked up the story and ran it as gospel truth.

The popular press of Edison's time made of him a superhuman figure, a scientific *Übermensch*. He was called necromancer, warlock, alchemist, immortal, king of the intellectual republic, Napoleon of inventors, emperor of science. And the titles of biographies (some written in his day and some much later) also indicate the stature he had in the popular imagination: the man who made the future, the greatest living man, the builder of civilization, benefactor of mankind, modern Olympian, the indispensable man.

A documentary first aired on November 18, 1980 (almost a hundred years after the perfection of the incandescent bulb), points out the two aspects of his personality the mass media still are drawn to. "The Wizard Who Spat on the Floor" makes of Edison both a mysterious, quasi-occult figure and a nuts-and-bolts self-made man. And understanding this dualism—in its real and mythic versions—helps to make sense of much of Edison's efforts.

His claim to be almost totally self-educated was based more on the anti-scholarly tenor of his time than on fact. He attended school until he was twelve years old, later taking every opportunity to point out how much he had achieved with little formal education. "School?" he'd say. "I've never been to school a day in my life! Do you think I would have amounted to anything if I had?" And while, as the years passed, his research organization used increasing numbers of academically trained scientists, he never let go of the resentment, fear, and deep uneasiness he felt around them. In 1875 he claimed to have discovered "etheric force" or the so-called Edison effect. Actually, what he'd stumbled onto were electromagnetic waves, a significant find. But even in the face of evidence from university physics professors, he insisted this was a new force in nature. The ridicule directed at him from the academic and scientific world further entrenched his distrust of what he called "the-o-ret-i-cal" science. He never forgot, nor forgave, them. It might be tempting to write off his decades-long enmity as mere professional combativeness, but there is more to the conflict than that.

Though Edison was instrumental in shaping much of modernism, he was at heart a primitive. Though called on by journalists to give his opinions on topics as diverse as immortality, war, religion, and politics, he was not a great thinker but a great doer. And this, too, puts him firmly in the tradition of science in the United States. The archetypal image of the American scientist is the doer, the hard-working exemplar of "Yankee know-how" and "can-do." At least until recently, Americans have depended on Europeans for theoretical science (often imported, like Einstein and Steinmetz).

Whereas university-trained scientists work first on the theoretical plane to determine the correct components for their experiments, Edison plunged in and worked endlessly to find the correct substances and combinations. His search for the best material to make incandescent bulb filaments is a good example. He tested thousands of organic fibers and sent explorers to South America and Asia to hunt down the perfect substance; he made innumerable adjustments and tests on the bulb before coming up with the correct combination. He frequently quoted the "99 percent perspiration and 1 percent inspiration" definition of genius, and had little tolerance for anyone who saw the situation differently. He was compelled to spend enormous amounts of time and money on pursuits that were sometimes utter failures—such as his ore-processing plant that ate up millions of dollars and ten years his life, and his futile search for an alternate source of rubber. Yet his belief in doggedness, work, and sheer strength of will never flagged.

The strain of the American puritan work ethic was deep in Edison, though lacking overt religious content. He truly did believe in salvation by work, though what exactly he was saving himself from is unclear. Perhaps he had spent so much time and effort inventing himself, building a public persona, that to veer away from it would threaten his sense of identity. Or it is possible, as is often the case with those shaped by the Calvinist ascetic world view, that no other way of seeing himself had the same authority or staying power. Or perhaps he was just a compulsive achiever, driven by a craving for fame and power.

In the United States, the traditional male response to crisis is abuse of alcohol and work. Edison was a

Edison in his laboratory.

teetotaler ("Prohibition is eternally correct.") But his obsession with work indicates a near hysterical undercurrent in his life. He needed, even well into his eighties, to be perceived as the great man of action, pushing back boundaries, defining himself against limits and barriers.

The popular literature of the day made much of this. He was in some ways the original Horatio Alger story. The authentic facts of Edison's life penetrated Alger's tales of self-reliance, hard work, and virtue, just as Alger's world view shaped Edison's view of himself.

Most biographies make much of Edison's days as a newspaper boy, teenaged printer, experimenter, and businessman. With little education, little support from his family, little besides his own "pluck and luck," he's shown again and again making his fortune. And as in the Alger stories of youths who make their way to success, Edison is depicted as not only hard-working but virtuous, almost pure. In Horatio Alger's fictional world though, it's not only perseverance and goodness that make for success. Luck, too, is important, and Young Tom, as he's called in the stories (though in real life he was Al or Alva) is shown having this in abundance too: almost burning down a barn, saving a child's life, being at the right place at the right time to "save" Wall Street during a stock-ticker breakdown. Though there's little emphasis on the more traditional masculine activities (hunting, sports, fighting, etc.), Edison is shown as a "manly" youth, depending on no one and willing to suffer greatly to succeed.

Throughout his life, much was made of Edison's alleged ability to go with little sleep. Again and again, he brought this up in interviews. He claimed that sleep was a sign of weakness, an acquired habit, something that cells, horses, and fish didn't need, so why should humans indulge in it? "In the old days, man went up and down with the sun. A million years from now we won't go to bed at all. Really sleep is an absurdity, a bad habit... we shall throw it off." According to Robert Conot's 1979 biography of Edison, this was all a "smokescreen, to divert attention from his habit of nodding off at work."[3]

Those who knew him well were aware of his ability to sleep just about anywhere—under benches and on tables in his lab, as well as at home. As the years passed and the legend of his ability to go without sleep became more a part of his public persona, some of his workers publicly declared their resentment and anger that he demanded that they sleep little yet he indulged himself and claimed that his superhuman ability remained. He was in fact known to sleep more than twenty-four hours at a time with only a brief break to eat.[4]

One of the most famous photographs of Edison was said to have been taken "as he appeared at five AM on June 16, 1888, after five days without sleep." He had called in many of his research people to make a "herculean" effort to complete changes on his phonograph—another one of his pseudo military campaigns to defeat a business rival. The popular press got wind of this story, and though they were barred from the laboratory, carried regular reports about the inventor's "frenzy" and the "orgy of toil" he

---

[3] Conot, 467.
[4] Wachhorst, 37.

and his men were enduring.

The photo shows Edison slouched over a table, listening to the finished photograph. One of his arms is outstretched in a posture of manly exhaustion. The set of his jaw was said to resemble that of a military victor at the end of grueling battle. An oil painting was made of the photo, and the artist endeavored to show Edison as the "Napoleon of Inventors." This image was then distributed widely as a poster advertising the Edison Phonograph Company.

The so-called five-day vigil had in fact lasted less than three, and had been very likely interrupted by frequent naps. But the effect of the stunt was exactly what he wanted: free publicity for his company and further "evidence" of his superhuman ability and will power.

Besides using Edison's biography as a moral and social exemplar, writers latched onto other aspects of the mythic persona for use in popular culture entertainment. In 1898, a sequel to H.G. Wells's *War of the Worlds* was serialized in the Hearst papers. Entitled *Edison's Conquest of Mars*, the novel makes the inventor a military hero, "America's genius of science" out for vengeance. The writer, Garret P. Serviss, has Edison investigate the Martian spaceships left after their defeat in Wells's novel. Edison unlocks the secrets of their technology and create space-faring vehicles as well as "a little implement which one could carry in his hand, but which was more powerful than any battleship,"[5] in other words, a death ray. The nations of earth join together and Edison leads

---

5   Garrett P. Serviss, *Edison's Conquest of Mars*, 9.

them into outer space. They arrive on Mars and the story then becomes a captivity narrative. Centuries before, the loathsome and corrupt Martians had come to earth and taken back with them an entire race of humans as slaves. Similar to Edgar Rice Burroughs's Mars and Venus tales, which were published not long after, Serviss's tale is as much a Western as it is science fiction. Eventually, Edison destroys the canals and puts an end to the despicable Martians.

Though Edison himself had nothing to do with the writing of the novel, the image of him streaking through space, wreaking vengeance on evil aliens, acting as the savior of civilization, must have pleased him. And though he might say publicly that he had little time for such frivolous pursuits, the image was consistent with the persona he felt compelled to project in other areas.

When he built his first research lab at Menlo Park—at that time a hamlet in rural New Jersey—it was ostensibly to get away from the distractions and demands of the metropolis. But the little world he created there had a distinctly communal air about it, like a religious retreat or hermitage.

The building itself (called the "tabernacle") was a plain two-story barn-like structure, a hundred feet long and thirty wide. Though often cited as the first facility in America devoted entirely to research, it resembled a meeting house or country church, with its tall windows and porch and front. Inside, there was even a pipe organ at one end of the first floor, for late-night musical gatherings.

This building was the "temple of science" and the

main room the holy of holies, where much of Edison's mythology was born. Here, Edison spent the best years of his life. Here, surrounded by acolytes and disciples, Edison perfected the incandescent light, the phonograph, and other ground-breaking inventions. The reverence that compelled Henry Ford to make his slavish recreation of the lab was not unique to him. Scientists, newspaper writers, ordinary neighbors, and curiosity seekers—all treated the place as though it were rich with supernatural power.

A week after publishing its announcement of Edison's successful testing of his lamp, the *New York Herald* ran an article overflowing with imagery that pushed the Edisonian myth into the realm of the grotesque.

> Invisible agencies are at his beck and call. He dwells in a cave and around it are skulls and skeletons, and strange phials filled with mystic fluids whereof he gives the inquirer to drink. He has a furnace and a cauldron and above him as he sits swings a quaint old silver lamp that lights up his long white hair and beard, the deep lined inscrutable face of the wizard, but shines strongest on the pages of the huge volume written in cabalistic characters... The furnace glows and small eerie spirits dance among the flames.[6]

Self-enclosed, in a tiny hamlet with little to offer except a simple boarding house for lodging, the tabernacle was a world unto itself. The only entertainment took

---

6   *New York Herald*, 29 Dec. 1879.

place in the lab, late at night. One of the men played a zither. The crew sang sentimental songs, told jokes and stories. Food was brought in. It was expected of all the men that they would forego pleasure, family life, even monetary reward, for their efforts.

Edison was fond of telling the story of a young man who came to Menlo Park hoping to work for him. He asked Edison what the hours were and how much he'd be paid. Though Edison was a millionaire many times over, his reply was, "We work all the time and we don't get paid."

After the first success with the photograph, Edison and his lab were the subject of intense attention by the press and general public. By carriage, wagon, and train, thousands of people came to see the lab, like the faithful believers on a pilgrimage to a high holy place. The Pennsylvania Railroad even organized excursions to bring scientists and curiosity seekers, the believers as well as the not-yet-converted to see this "Vatican of Science," the "Nineteenth Century Miracle."

Edison was at first delighted to show off his laboratory and amuse the visitors with his inventions. He downplayed the title "Wizard," yet took frequent opportunities to amaze, baffle, and frighten those who came into his presence. As stated by W.K.L. Dickson, who had come from England to seek work with Edison, "A species of glorified mist soon enveloped" Edison, and he was "regarded with a kind of uncanny fascination, similar to that inspired by Dr. Faustus of old; no feat would have been considered too great for his occult attainment. Had the skies been suddenly darkened by a flotilla of airships" carrying "a deputation of Martians, the phenomena would have been accepted as a proper achievement of the scientist's genius."[7]

Rumors circulated among his rural neighbors that he had created a device that allowed him to hear them talking miles away. Windows glowed far into the night and shadowy figures were seen moving about the fields near the labs, carrying strange lights.

It was not just the Faustian seeker of knowledge that Edison was compared to but also Aladdin, Mephistopheles, Frankenstein, Prometheus, Merlin, and even Satan. He was credited with baleful and mysterious powers, even the ability to rewrite the laws of nature. Newspapers refer to his journals as the "Edisonian Book of Genesis."

Magic and science blur not only in the popular mind but are actually linked at a deep ideological level. Ioan Culiano's *Eros and Magic in the Renaissance* makes a strong case that magic and science "in the last analysis represent needs of the imagination, and the transition from a society dominated by magic to a predominantly scientific society is explicable primarily by a change in the imaginary." Regarding mankind's emotional, as well as intellectual, needs, he goes on to explain that the origins of modern science "could not have appeared without the existence of factors able to cause modification of

---

[7] Dickson and Dickson, *Life and Inventions of Thomas Alva Edison*, 101.

man's imagination."[8] Edison and his amazing creations became a focal point for America to project its fantasies onto, a lightning rod to attract and channel the cultural forces of the day.

Harper's Weekly describes Edison as both a Renaissance magician and the paragon of modernity:

> A single gas flame flickers at one end of the long room, and turned pieces of wood, curious shapes of brass, and a wilderness of wires, some straight, others coiled and spiral and kinked, the ends pinched under thumbscrews, or hidden in dirty jars, or hanging free from invisible supports—an indiscriminate, shadowy, uncanny foreground. Picking his way circumspectly around a bluish, half-translucent bulwark of jars filled with azure liquid, and chained together by wires, a new picture meets [the visitor's] bewildered eyes. At an open red brick chimney, fitfully outlined from the darkness by the light of fiercely smoking lamps, stands a roughly clothed gray-haired man, his tall form stooping under a wooden hood which seems to confine noxious gases and compel them into the flue. He is intent upon a complex arrangement of brass and iron and copper wire, assisted by magnets and vitriol jars, vials labeled in chemical formulae, and retorts in which to form new liquid combinations. His eager countenance is lighted up by the yellow glare of the unsteady lamps, as he glances into a heavy old book there, while his broad shoulders keep

---

[8] Ioan Culianu, *Eros and Magic in the Renaissance*, xviii-xix.

out the gloom that lurks in all corners and hides among the masses of machinery. He is a fit occupant for this weird scene; a midnight workman with supernal forces whose mysterious phenomena have taught men their largest idea of elemental powers: a modern alchemist, who finds the philosopher's stone to be made of carbon, and with his magnetic wand changes everyday knowledge into the pure gold of new applications and original uses. He is Thomas A. Edison, at work in his laboratory, deep in his conjuring of Nature while the world sleeps.[9]

As Mircea Eliade writes in *The Forge and the Crucible*: "The ideology of the new epoch, crystallized around the myth of infinite progress and boosted by the experimental sciences and the progress of industrialization which dominated and inspired the whole of the nineteenth century, takes up and carries forward—despite its radical secularization—the millennial dream of the alchemist."[10]

The word "alchemist" appears frequently in descriptions of Edison, and while the popular press used this term more to connote strangeness and otherworldly power, they hit on an important element in the Edison myth. Mircea Eliade's book points out the connection between the alchemist and the "sacred smith," between the "master of fire" and the holy technician.

Edison in some ways fits the archetype of the

---

[9] "Edison in his Workshop," *Harper's Weekly*, 2 Aug. 1879: 17-19.

[10] Mircea Eliade, *The Forge and the Crucible*, 172, 173.

alchemist—converting base metals to the "gold" of useful technology, using arcane science to transform himself into a new, perfected man. But it is as the primal metal-worker that he had the greatest impact on his culture. Eliade describes dozens of mythic and folkloric instances where the smith is the agent of the gods, bringing civilization to humanity. Asian, Norse, African, and South American stories tell of a demigod metal-worker who "is believed to have brought with him the tools necessary for the cultivation of the soil and thereby became a 'civilizing hero.'" Just as Edison was both admired and shunned, celebrated as a hero of progress and feared for his powers, so the primal smith is a figure of great ambivalence.

"The common origins of the sacredness of shamans and smiths is shown in their 'mastery over fire.' In theoretical terms, this 'mastery' signifies the attainment of a state superior to the human condition. What is more, it is the smith who creates weapons for heroes. It is not their material creation that matters but the magic with which they are invested; the smith's mysterious art transforms them into magic tools."

Edison's ten-year campaign to perfect his iron-ore-crushing system fits neatly in with the metal-taming myth. Off in a wilderness area without family, surrounded only by a society of like-minded males, he was—though well into his sixties—like a youth suffering to gain mastery over metal, and thus, himself. In other ways, too, the archetypal shamanic initiation process shares many basic characteristics with the legendary process by which Edison achieved his inventive victories.

He spent long periods with little food, shelter, or sleep. He was plagued by fear and fatigue, hunger and danger, which created a visionary state of receptivity (for the shaman-initiate a doorway to the divine, for Edison access to inventive intuition).

Eliade argues persuasively that the essence of the alchemical dynamic is the elimination of the "interval of time which separates the present condition of the 'imperfect' (crude) metal from its final condition (when it would become gold)." The philosopher's stone "achieved transmutation almost instantaneously; it superseded Time."

In an analogous manner, Edison's most successful inventions were those that exploited the dynamic of extension. They achieved a similar result as the alchemical process: in effect folding space and time so that points once far distant could become adjacent. The telephone allows individuals to speak to one another over huge distances. Once the phonograph and motion pictures were perfected, living images could transcend time. Once the system of electrical transmission was functioning, mechanical power could be converted, then extended thousands of miles, transcending space. "The visionary's myth of perfection, or more accurately, of the redemption of Nature, survives in camouflaged form in the pathetic programme of the industrialized societies, whose aim is the total transmutation of Nature, its transformation into 'energy.'"[11]

Though as time passed and the term "wizard" came to be replaced by "genius" and "Great American," the

---

11   Eliade, 78, 85, 90, 173.

concept of the wizard remains useful for understanding the man.

The image of the wizard encapsulates two halves of the true Edison: the combative warrior-of-science and the primitive, almost naïve self drawn to bizarre, "non-scientific sciences" such as spiritualism, patent medicines and pseudo-Eastern religion. And while it might seem that there is a contradiction between the cold, rationalist scientist and a believer in mystical transcendence, in fact these go hand in hand. Edison's work was not just about the elevation of Western technology to new heights, but about the elevation of the inventor himself. As Eliade notes, the alchemist "prolongs and consummates a very old dream of *homo faber*: collaboration in the perfecting of matter while at the same time securing perfection for himself." Edison's efforts must be seen in the context of the cultural beliefs they rose from. He was a man of faith, and "it is in this faith in experimental science and grandiose industrial projects that we must look for the alchemist's dreams. Alchemy has bequeathed much more to the modern world than a rudimentary chemistry; it has left us its faith in the transmutation of Nature and its ambition to control Time."[12]

Culiano further develops this idea:

> Historians have been wrong in concluding that magic disappeared with the advent of 'quantitative science.' The latter has simply substituted itself for a part of magic while extending its dreams and its goals by means of

---

12 Eliade, 169, 174.

technology. Electricity, rapid transport, radio and television, the airplane, and the computer have merely carried into effect the promises first formulated by magic, resulting from the supernatural processes of the magician: to produce light, to move instantaneously from one point in space to another, to communicate with faraway regions of space, to fly through the air, and to have an infallible memory at one's disposal. Technology, it can be said, is a democratic magic that allows everyone to enjoy the extraordinary capabilities of which the magician used to boast.[13]

One chapter in Edison's life that is almost universally absent from his biographies is his involvement with Theosophy. Founded in 1875, in New York, by Madame Helena Blavatsky, the Theosophical Society took elements from Hinduism, Buddhism, and American Spiritualism and combined them into a unified system of belief. Theosophists were particularly interested in developing supernatural faculties they thought latent in humans: levitation, materialization of spirits, psychic communication. Blavatsky's *Isis Unveiled* (published in 1877) and *Secret Doctrine* (1888) both had a major influence on occult and mediumistic thinkers of her day.

Edison met H.S. Olcott (Blavatsky's right-hand man) in Paris and they spoke at length about "occult forces." Olcott visited Menlo Park a number of times in 1878, and on April 4 Edison signed his membership "diploma" and pledge of secrecy, which reads, "In accepting

---

13   Culianu, 104.

fellowship with the above mentioned society, I hereby promise to ever maintain ABSOLUTE SECRECY respecting its proceedings, including its investigations and experiments, except in so far as publication may be authorized by the society or council, and I hereby PLEDGE MY WORD OF HONOR for the strict observance of this covenant."[14]

Though Edison denied later in life that he had had anything to do with Theosophy (another of his many revisionist biographical "facts"), he was continually attracted to mysticism and the theory of memory, soul, and intelligence that he developed was strongly influenced by Blavatsky's ideas. Memory and mind, Edison claimed, consist of microscopic particles of matter. These "entities" are able to travel outside the body, and even—he theorized—may have come from outer space, bringing to earth knowledge of other planets. Gathering like a swarm of bees, the entities take up residence in the human brain and thus create intelligence. Edison claimed that these "little people," as he called them, could also explain heredity and instinct. At times the swarms, like rival armies, came into conflict within a person. As Edison wrote in his diary,

> They fight out their differences, and then the stronger group takes charge. If the minority is to be disciplined and to conform, there is harmony. But the minorities sometimes say "to hell with this place, let's

---

[14] Sylvia Cranston, *HPB: The Extraordinary Life and Influence of Helena Blavatsky*, 184.

get out of it." They refuse to do their appointed work in a man's body, he sickens and dies, and the minority gets out, as does too, of course, the majority. They are all set free to seek new experience somewhere else.[15]

Death, according to Edison, is merely the departure of the entities from the human body.

Late in life, Edison used photographic plates in experiments, trying to communicate with the memory swarms. He also claimed to have sold certain telegraphic patents, in 1885, to a spiritualist, who hoped to use the technology to listen to voices from another world. The parapsychologist Bert Reese was introduced to Edison by Henry Ford and in a number of laboratory experiments convinced Edison of his psychic powers. Meeting Houdini, Edison attempted to replicate Reese's telepathic results, winding electrical coils around his own head and around those of three others. Houdini was not convinced.

In the early 1920s, interviews were published in which Edison claimed to be working on technology with which he could communicate with the dead. Whether he actually did work on the device is questionable, but he spoke at great length about the possibility that science could solve the mystery of life after death. "I have been at work for some time building an apparatus to see if it is possible for personalities which have left this earth to communicate with us. If this ever is accomplished, it will be accomplished not by occult, mystifying, mysterious or weird means, such as are employed by so-called mediums, but

---

15 Thomas A. Edison, *Observations*, 212-13.

by scientific methods."[16] Though the methods and results might have been identical, Edison made a distinction between science and occultism. No matter how ludicrous the experiment, as long as it was couched in terms of reason and objectivity, it was potentially acceptable to him.

Edison, like most great individuals, was a tangle of contradictions. He could be as guileless as a child and as crafty and manipulative as a sideshow barker. He's credited with being the greatest inventor in his country's history, yet he had only the most rudimentary understanding of chemistry, physics, and mathematics. He had more patents registered under his name than any other American, but could publicly boast, "Everybody steals in commerce and industry. I've stolen a lot myself. But I know how to steal."[17] He's described as the "greatest living man" and the "benefactor of mankind" yet was vindictive, cruel and "waged war" against his scientific and commercial enemies.

Throughout Edison's entire adult life, he felt the need to define himself by confrontation. With little schooling and little ability to understand upper-level science, he attacked education as worthless or even injurious to young people. Contemptuous of those who trafficked with mediums and occult ideas, he created an entire pseudoscientific theory of life after death. Heaped with medals, honors, and awards, he continued to present himself as an ordinary man who just worked

---

16   B.C. Forbes, "Edison Working on How to Communicate with the Next World," *American Magazine*, Oct. 1920: 10, 11, 82, 85.

17   Matthew Josephson, *Edison: A Biography*, 180.

hard and allowed himself no time for frivolity. Accused of patent infringement and claiming his underlings' work as his own, he instigated dozens of legal suits and attacked the "money men" of Wall Street for taking advantage of him.

Business endeavors and publicity stunts alike were framed by a warlike worldview. The quest for the perfect light bulb filament, during which he sent his scientific conquistadors into the jungles of Asia and South America, is a clear example of his belligerent mentality. In his world view, there was a perpetual conflict between science and nature. Edison relished the Napoleonic comparisons. He was frequently described as "subduing nature" and "wresting secrets from her bosom." One reads of his conquering, mastering, controlling, subjugating, the natural world.

His first major "war" took place when Edison was still quite young. Western Union and the Automatic Telegraph Company were competing with each other for control of the very lucrative communication market. Edison was brought in to adapt certain technologies so that patent restrictions could be evaded. He was in effect a scientific mercenary, working for whichever side would pay him the most.

After the 1876 Philadelphia Centennial Exposition, where Bell's telephone made a huge public impact, Western Union called in Edison to make improvements in its technology so that they could cut into the market. Again Edison was embroiled in a patent war as he pitted himself against Alexander Graham Bell in the race to perfect the telephone. Edison eagerly jumped into the battle, both because he was genuinely interested in the scientific challenge and because it was another opportunity to define himself through conflict.

Prior to the Great War of the Currents, the best example of Edison's combative streak was his campaign to unseat the gas companies as the main source of illumination. Gas lighting, by the late 1870s, had become a major industry in the United States, with annual revenues of over $150 million.[18] Other electrical engineers were already in the field, but once Edison had decided to commit himself to the battle, he became the major player. As soon as his plan was announced, journalists were eager to cover the story. A reporter from a *New York Tribune* said to him, "If you can replace gas lights, you can easily make a great fortune."

Edison's reply was, "I don't care so much about making my fortune as I do for getting ahead of the other fellows."[19]

Matthew Josephson describes the dynamic this way: "To have more money meant little; he had enough, he said, for his needs. But to stand as a leader among the world's foremost inventors, to make again and again a great impact on society and industry—even to 'change the world' if possible—meant everything."[20] Before he focused his attention on the specific technological problems of cutting in on the gas companies' turf, Edison analyzed the societal and commercial environment where he would

---

18   Josephson, 180.
19   Qtd. in Josephson, 180-81.
20   Josephson, 181.

first attack. Then he gathered dozens of books on gas illumination and all the back issues he could find of the gas industry journals. He studied the seasonal changes in consumption, the logistical operations, the geography and organization patterns. He drew charts and tables, then mapped out his strategy. By the end of his preparatory phase he knew more about the gas industry than almost anyone in the world.

There were always two sides to Edison's campaigns: the actual implementation of the new technology and the public relations efforts. The Philadelphia Exposition marked the first appearance of the "Edison Darkey," a tall black man with an incandescent light attached to his derby. Wired through his coat and pants to copper plates on the heels of his shoes, he would dance on a floor laid with conductive strips. When his shoes made contact, the lamp came on, bathing the spectators and brilliant light. The "Darkey" appeared at many expositions and fairs throughout the middle 1880s, a sign of Edison's increasing presence in American life.

Often implementation and publicity shaded into each other. Once the actual campaign of wiring a section of lower Manhattan had begun, Edison was literally in the trenches with his men. He supervised the digging and laying of electrical mains, sleeping in filth and struggling with conduit alongside common laborers just as mythic military leaders allowed themselves to suffer the same privations as their lowest foot soldiers.

He was just as willing to embrace grandiose and flamboyant displays if that would further his cause. In

"Wired, but human," "The Edison Electrical Darkey".
Scientific American, Oct. 18, 1884.

the fall of 1884, like a triumphant army taking possession of a conquered city, a military-style parade advanced down Fifth Avenue in New York. Hundreds of men, all of them Edison employees, marched in a hollow square formation, each wearing a helmet topped by a small glow lamp. In the square were a steam engine and an Edison dynamo on wheels. The men were attached by wires to the dynamo. At the front of the column was a gaudily dressed commander on a warhorse, carrying a baton with light bulbs at each end. The rapidly developing technology and Edison's doggedness had combined to bring the gas companies to their knees, and the incandescent bulb became the symbol of Edison's victory.

By the end of the decade, though, he found himself on the other side of the David and Goliath model. As his mythos crystallized and his inventive powers waned, Edison became an increasingly conservative presence. Public expectation demanded that he continued to succeed, and yet other scientists were making rapid progress in the fields where he'd once been preeminent. Others—notably Nikola Tesla—were advancing technologies that would quickly make much of Edison's work obsolete. By the late 1880s, Edison found himself on the defensive, in the same position as the gas companies only a decade before. The Great War of the Currents was soon to threaten not only Edison's technological empire, but his purposes and even his sense of self. His unique synthesis of naïveté and guile, combativeness and apparent altruism, would make him a formidable opponent, willing to use any tactic necessary to maintain his dominant position.

The New-York Times.

June 7, 1889

OFFERS HIMSELF FOR EXECUTION

A curious letter has been received by Superintendent Lathrop from Philadelphia, signed 'A.Z.' It states that the writer is a man "down on his luck" and seeing that the prison authorities in New York are still somewhat doubtful as to the efficacy of the new electrical apparatus for the execution of criminals, he offers himself as a subject for experimentation, if the State will promise to pay his wife the sum of $5000. The letter is evidently written by a man of education, and in all sincerity. A. Z. says in conclusion that a personal in the *Philadelphia Ledger* accepting the proposal will bring him to Albany at once to make the final arrangements with authorities.

Chapter Four

THE TRANSFORMER

If Thomas Edison was the archetypal inventor, then Nikola Tesla was the ultimate scientific outsider. Whereas Edison was one of the most famous men of his day, Tesla's fame peaked early and by the time of his death in 1943, he'd slid into obscurity. Edison became a millionaire many times over; Tesla died a pauper. Though much of his work quickly became obsolete, Edison was seen as an American success story. Tesla—so far ahead of his time that physicists and engineers are still trying to catch up with him—was plagued by failure and frustration.

In the popular imagination Edison and Tesla stood poised against each other, dueling like demigods on some higher plane with weapons no mere mortal could hope to comprehend. More so even than his former employer, mentor, and hero, Tesla merged the imaginary and the actual. Bathing himself in huge quantities of electrical energy, making large buildings quake with a tiny vibrating device, shimmering in a high frequency halo, Tesla brought the real and unreal closer than any other person of his age.

His magnum opus was the perfection of the alternating current system. Years before, Edison had broken the ground, transmuting electricity from a scientific curiosity to an everyday reality. But it was Tesla who took electricity to the next, and all-important, step. Edison's work was based on direct current, in which the flow of electrons moves in only one direction. But electricity by its nature moves back and forth, like a river flowing upstream and down, switching directions many times a second. Tesla understood the vast power inherent in alternating current and from the very beginning of his work as an engineer he pursued the idea that motors, sources of light and heat, would one day run on AC rather than the then-dominant DC.

Unlike Edison, who labored doggedly, making thousands of experimental attempts before perfecting an invention, Tesla worked largely in his mind. He intuited, visualized, and designed most of his inventions without ever committing them to paper. Sometimes, as in the case of the first AC motor and dynamo, the actual construction of the device didn't take place until many years had passed since its conception. A man of enormous memory, concentration, and cognitive power, Tesla's most far-reaching creation came to him after a long bizarre illness that brought him close to death.

In 1881, Tesla left his native Croatia (he was of Serbian parentage) to work as an engineer in the American Telephone Company's newly opened Budapest exchange. Still a young man, he had been there only a year when he was stricken with a condition that neither he nor his doctors could explain. In lay terms, his illness might be called a nervous breakdown, but as with everything in Tesla's life, the illness didn't fit neatly into any standard category. His pulse raced to 260 beats per minute. His body twitched and trembled continuously. His doctor had little to offer him except large doses of potassium bromide.

The most peculiar symptom, however, was the hyperacuteness of all his senses. In Tesla's autobiography, he describes the torture which sight, sound, and touch became for him:

> A carriage passing at a distance of a few miles fairly shook my whole body. The whistle of a locomotive twenty or thirty miles away made the bench or chair on which I sat vibrate so strongly that the pain was unbearable. The ground under my feet trembled continuously. I had to support my bed on rubber cushions to get any rest at all. The roaring noises from near and far often produced the effect of spoken words which would have frightened me had I not been able to resolve them into their accidental components. The sun's rays, when periodically intercepted, would cause blows of such force on my brain that they would stun me. I had to summon all my willpower to pass under a bridge or other structure as I experienced a crushing pressure on the skull.[1]

Though these symptoms might be regarded as

---

1   Nikola Tesla, *My Inventions*, 60.

Nikola Tesla sending 250,000 volts through his body, 1894 engraving from *The World*.

fantasy or the hallucinations of a young man pushed over the edge by overwork and high expectations placed on himself, they echo the compulsions and repulsions that later came to rule Tesla's life. Food, sex, physical contact of any kind, became extremely problematic for him. On some level, he was at war with his body, with the human form itself. Later on in life, in order to protect himself from the results of his neurotic dreads and drives, he developed a theory that all humans are automatons, or "meat machines" as he called them. Living largely in his head, the limitations and indignities of being a creature of flesh and blood caused him no end of discomfort. Regarding his illness, he wrote: "It is my eternal regret that I was not under the observation of experts in physiology and psychology at that time. I clung desperately to life, but never expected to recover."[2]

Like the traditional initiation journey of the shaman, Tesla's illness tormented him physically and tore him apart psychologically. His condition brought him close to death—or so he thought—and left him a new man, more receptive to the visions, the until-then-hidden understanding that he would soon use. And like the shaman, he brought back from "the other side" knowledge that was not just for his own gain, but which would benefit all humankind. He was convinced—and rightly so—that the inventions born in his delirium would revolutionize the way humans lived.

He had, while still in school in Karlovac, Croatia, become obsessed with the idea that direct current would

2   Tesla, 60.

one day be superseded by alternating current. Because AC was thought dangerous and impossible to tame, it was almost always converted to DC before use. A commutator was used in DC technology to limit the flow to one direction. It was in essence a series of wire brushes that rubbed against the rotating elements of the generator. DC was used by Edison in all his lighting systems. But not only did direct current dynamos have many more moving parts that broke down quickly, inherent in DC was the problem that its power falls off severely if transported more than a few hundred yards.

At the Austrian Polytechnic School at Graz, Tesla was publicly humiliated by an instructor for his belief in the superiority of AC; however, Tesla was determined to solve the problems that made it, up to that time, uncontrollable and unusable.

> When I undertook the task it was not with a resolve such as men often make. With me it was a sacred vow, a question of life and death. I knew that I would perish if I failed... A thousand secrets of nature which I might have stumbled upon accidentally I would have given for that one which I wrested from her against all odds and at the peril of my existence.[3]

Clearly this was not merely scientific curiosity for Tesla. Nor was he driven by greed or the desire to exalt himself over others. It was an issue of self-definition, of identity and meaning. Like the mythic philosopher who

---

3   Tesla, 60-61.

trades all for a glimpse of ultimate wisdom, Tesla devoted himself, sacrificed, and struggled so that alternating current could be harnessed.

Soon after recovering from his breakdown, he was walking in a park with his friend Anital Szigety. The sun was just going down and the brilliant colors reminded him of a passage from Goethe's *Faust*. He recited to his friend from memory:

> The sun shudders and shimmers,
> > having survived the day,
> > then moves on to stir new life.
> Oh, that no wing's there to lift me
> > from the ground to reach, to follow.
> But as a beautiful dream, it's gone.
> Oh, the spirit-wings fade, and no wings
> > of flesh come to take their place.[4]

Overcome by a mixture of loss and melancholic elation, Tesla became motionless, as if in a trance. "The idea came like a flash of lightning and in an instant the truth was revealed."[5] Immersed in nature's beauty, in the pathos and ecstasy of Faust's quest for knowledge, Tesla intuited in one sudden leap the method by which alternating current could be tamed. He drew with a stick in the earth the systems he would present six years later when he lectured before the American Institute of Electrical Engineers.

---

4   Author's translation.
5   Tesla, 61.

Looking directly into the flames of the dying sun, Tesla found the model that allowed him to make his breakthrough. At that time, it was thought that the sun's light was created by the vibrations of the molecules of its outer gaseous layer. Making an intuitive leap—though one based on erroneous theory—he generalized from this to an overarching conception of all nature operating on the principle of vibration, an analog to the cyclic quality of alternating current.

For Tesla, this wasn't merely the revelation of a new idea but an intense, almost spiritual, event. The vision was palpable, overpowering. He saw the alternating current system in all its detail as though it were really there, floating before him. The ability to visualize with almost supernatural intensity was something that he carried from childhood to his death. In an article published in the *American Magazine* of April 1921, he explains:

> During my boyhood, I had suffered from a peculiar affliction due to the appearance of images, which were often accompanied by strong flashes of light. When a word was spoken, the image of the object designated would present itself so vividly to my vision that I could not tell whether it was real or not... Even though I reached out and passed my hand through it, the image would remain fixed in space.
>
> In trying to free myself from these tormenting appearances, I tried to concentrate my thoughts on some peaceful, quieting scene I had witnessed. This would give me momentary relief; but when I had

done it two or three times the remedy would begin to lose its force. Then I began to take mental excursions beyond the small world of my actual knowledge. Day and night, in imagination, I went on journeys—saw new places, cities, countries, all the time I tried hard to make these imaginary things very sharp and clear in my mind. I imagined myself living in countries I had never seen, and I made imaginary friends, who were very dear to me and really seemed alive.

This I did constantly until I was seventeen, when my thoughts turned seriously to invention. Then, to my delight, I found I could visualize with the greatest facility. I needed no models, drawings, or experiments. I could picture them all in my mind.[6]

This state of visionary ecstasy that Tesla experienced in the park in Budapest was not to last long however. He had discovered the principle that would allow him to harness alternating current, but implementing his idea would require financial backing.

Shortly after his vision, the telephone facility where he worked was sold. His supervisor wrote him a letter of recommendation and in 1882 Tesla was employed in Paris as a troubleshooter for the Continental Edison Company, which installed lighting systems and electric motors and dynamos under Edison's patents. Tesla traveled to various facilities in France and Germany, investigating engineering problems and doing needed repair. He quickly made suggestions for improvements in the

---

6 Qtd. in John O'Neill, *Prodigal Genius*, 256.

company's dynamos and was allowed to apply his ideas to certain of the equipment. These changes were successful and he was then asked to design automatic regulators, which too made marked improvements in the systems.

While working in Paris between 1882 and 1884, Tesla met many American electrical engineers and tried to interest anyone who would listen in his grand plans for alternating current. But the visionary nature of his work, and the brooding presence of Edison, who stated repeatedly and categorically that direct current was the only safe and efficient system to use, made the possibility of Tesla's ideas being taken seriously very slim. Continually rebuffed by the executive at Continental Edison, Tesla decided to demonstrate his new system by himself. In Strasbourg, Tesla constructed without benefit of blueprints or plans an AC motor and dynamo. The mayor of Strasbourg was interested and gathered a number of wealthy potential backers to witness the test runs. The system worked exactly as Tesla had planned, but none of the men gathered there showed any interest in it. Returning to Paris, Tesla was—not for the last time—cheated out of money due to him from the Edison organization. None of the supervisors he'd worked with would take responsibility for paying him the bonus he had been promised for his improvements on the Edison regulators. Disgusted, Tesla resigned and swore not to show any of his successful AC equipment to the Edison people in Europe.

However, one of the administrators of the company was Charles Batchelor, an engineer who had worked with Edison when the inventor made his alterations on Alexander Graham Bell's first telephone. Not only a business associate, Batchelor was also at that time one of Edison's best friends. He urged Tesla to go to the United States to work with Edison. "I know two great men and you are one of them; the other is this young man," Batchelor wrote in his letter of recommendation. Tesla, who held Edison as his model and hero, took the opportunity. He sold all his belongings, bought a ticket for the transatlantic trip on the *Saturnia*, and arrived in New York in June of 1884.

Tesla exited the Castle Garden immigration facility with—by his own account—only four cents in his pocket, along with a few poems, some calculations, and a drawing of a flying machine. The next day he walked to Edison's headquarters in lower Manhattan. Though they had met briefly in Europe, it's unlikely that Edison remembered the young inventor. Tesla arrived at Edison's plant on South Fifth Avenue (now West Broadway) and presented his letter of recommendation.

The meeting of these two men in the dirt and noise of Edison's shop might be seen as the first skirmish in what became the AC-DC war. On almost every level they were opposites. Tesla was a man of science and sophistication, with upper level degrees and fluency in five languages. Edison was a homespun anti-intellectual. Tesla was a representative of European culture; Edison was a crude, sentimental American. At six foot six, Tesla towered over Edison. And weighing less than 140 pounds, he contrasted sharply with Edison, who by this time was heavy and solid as a tree stump. Tesla's hair was straight, shiny and black; Edison's was prematurely gray. Tesla's head was a wedge, with the chin coming to a sharp point. Edison's head was described by the popular press as massive and full. Even on the level of personal hygiene they were hugely different: Tesla fastidious to the point of obsession and Edison happy to be covered with dirt and grime.

Most important though, Edison was now the voice of conservatism and his young visitor was a revolutionary. Tesla might have represented the old world in his manners and personal habits, but his vision of science made Edison seem archaic, a hopeless throwback, in comparison.

When Tesla described his discoveries to Edison and declared that AC was the inevitable wave of the future, Edison told him brusquely that he wasn't interested. There was no point in pursuing it, Edison said, and anyone who thought otherwise was a fool.

Years later, when the dust and noise of the AC-DC war had settled, there was still a strong undercurrent of conflict and resentment between the two men. Tesla made a clear distinction between the discoverer of scientific principles and the mere inventor. Tesla saw himself as a pioneer into new realms of knowledge, a prophet and high priest devoted to learning, whereas Edison was simply an exploiter of that knowledge. When talk of their sharing the Nobel Prize in physics reached Tesla, he said he would decline (the first person to do so) because placing him in the same category as Edison would obscure if not eliminate the distinction between the discoverer and the inventor.

Edison, meeting Tesla on his home turf, sensed this haughty attitude immediately. Nonetheless, because of Batchelor's letter of high praise, Tesla was hired onto Edison's staff.

To put him in his place, Edison assigned Tesla to tasks far below his abilities. But when a few weeks later Edison needed an engineer to make repairs on the steamship *Oregon*'s electrical system, he sent Tesla. The ship's date of departure had already passed and the owners were frantic to get the *Oregon* out to sea. Pressing members of the crew into service, Tesla worked nonstop from early afternoon until four in the morning, tracking down shorts and getting the dynamos back in operation.

Edison was impressed with Tesla's abilities and passion for hard work. Tesla rose quickly through the ranks and was soon involved in design and operations. Tesla observed numerous ways in which the Edison equipment could be made more efficient and less costly to operate. After listening to Tesla outline his plan, and much impressed by the claims of savings, Edison offhandedly said, "There's fifty thousand dollars in it for you if you can do it."[7] Tesla was unused to Edison's ways and took him at his word. He began immediately, working eighteen-hour days to redesign two dozen types of dynamos, replacing long-core magnets with shorter ones, and creating a series of automatic controls. Months later, after prototypes of the new equipment had been built and found to

---

7   O'Neill, 64.

exceed Tesla's promised results, he went to Edison and asked for the $50,000.

"Tesla, you don't understand our American humor," was Edison's reply. No money or further explanation was forthcoming. Shocked to be the butt of one of Edison's practical jokes, appalled that he'd been duped into working enormous amounts of overtime to save the organization thousands of dollars with no compensation or even thanks, Tesla resigned immediately.

He had worked for Edison less than a year. But during that time he'd developed a reputation within electrical engineering circles. A group of investors soon offered him capital to form a company under his own name. He grabbed the chance, hoping to finally put his alternating current ideas into practice. The Tesla Electric Light Company was established, with headquarters in Rahway, New Jersey. His backers, however, were only interested in the development of DC arc lighting for factories and street use.

Within a year Tesla had taken out patents and begun the manufacture of the desired equipment. The system was first put into use on the streets of Rahway. Again, Tesla's naïveté and unfamiliarity with American ways of business brought him to disaster. He was paid little by the investors and given stock certificates that he found were near-worthless when he tried to cash them. Tesla was out of work again, with almost nothing to show for his efforts.

At this point, the economic slump of the 1880s had developed into a full-blown depression. Tesla was unemployed, with few options and little chance of ever putting his revolutionary ideas into effect. From the spring of 1886 to the spring of the following year, he worked at menial electrical repair jobs and even as a ditch digger for two dollars a day. He'd traveled halfway around the world, worked with some of the finest minds of his day, made important discoveries, and yet found himself breaking the frozen dirt with a pick ax: filthy, hungry, dejected. During the winter of 1887, however, he came to the attention of the foreman of the work gang, who had also been forced by the slump to work well below his abilities and educational level. The foreman was impressed with Tesla's engineering knowledge and put him in contact with Alfred K. Brown, a manager with the Western Union Telegraph Company. Brown was fascinated by Tesla's ideas and, just as important, was also willing to put up money to create a new company in Tesla's name.

Brown and Tesla were joined by lawyer Charles Peck and Anital Szigety—Tesla's friend from his Budapest days—to form the Tesla Electric Company. In April of 1887 they opened a laboratory at 33-35 South Fifth Avenue, near Bleecker Street, within sight of Edison's main New York facility.

Work began immediately at the small factory and soon they produced complete systems of AC equipment: dynamos for generating current, motors, automatic controls, and transformers. The systems created in the lab were identical to those Tesla envisioned in the park in Budapest. Five years had passed since his revelation, and though he had drawn no plans, committed nothing to paper, he remembered the design perfectly. In fact, it wasn't just one system he had carried in his mind for those years, but three distinct sets of technology—for single, double, and triple-phase currents. He submitted the double-phase motor to William A. Anthony, a professor of electrical engineering at Cornell. Tests showed that the motor was equal in efficiency to any direct current motor.

On October 12, 1887, Tesla's attorneys applied for a patent covering the AC system as a whole. The United States Patent Office refused to consider such an "omnibus" application. Tesla then broke the systems into seven distinct inventions and applied for the patents in two groups, in November and December. Six months later, the application ascertained utterly original and without any remotely similar competition, the patents were granted.

Professor Anthony and Thomas C. Martin—of the American Institute of Electrical Engineering, and editor of *Electrical World*—persuaded Tesla to give a lecture to his peers. On May 16, 1888, he appeared before the AIEE. The date was a landmark for Tesla and the electrical community. In the lecture he explained clearly the design and theory behind his revolutionary creations.

The energy of an electrical current is measured primarily by two qualities: the amount of current flowing (amperage) and the pressure or force behind the current (voltage). All substances that electricity flows through have an innate resistance to that flow, a tendency to impede the movement of electrons. The amount of energy lost (converted to heat) due to resistance is the same whether the pressure behind is tiny or enormous. Thus, it was of great importance that scientists find a way of transporting electrical energy at very high voltages.

In the direct current system typified by Edison's apparatus there is no practical way of increasing or decreasing the amount of electrical current once it leaves the generator. And because in the DC system there is a progressive loss of power the farther one gets from the dynamo, voltages vary greatly more than a half-mile from the generating plant. Also, the amount of copper needed for the wiring is far greater in the Edison system, adding considerably to the cost. As a result, the Edison generating scheme was by nature limited to very small areas. In order to supply a large metropolitan area with power, dozens of the generating facilities would be necessary. And beyond the urban areas, the possibility of efficient distribution of power was even slimmer.

Tesla's AC system made it possible to transport electrical power over hundreds of miles at very high voltages, and then step the power down through the use of transformers, for use in homes and manufacturing. Thus, using Tesla's system, a very small wire could carry thousands of times more energy than it could using DC.

The fundamental principle of Tesla's discovery was the rotating magnetic field. Producing two or more alternating currents out of phase with each other, he created in effect a magnetic vortex turning in space. Attempts had been made before this time to create an AC motor, but because the magnetic fields created by AC reverse themselves many times per second (as the current itself changes direction), a great deal of useless vibration was churned up, making the efficiency of the motors quite low. Tesla overcame this vibration problem by the creation of his controlled magnetic cyclones. An added benefit was the elimination of the commutators and brushes needed for the passage of current in the DC motor.

In the Tesla polyphase system, two or more circuits, each carrying the same frequency of current, are deployed out of step with each other. One way to visualize this is to imagine a gasoline engine with only one cylinder. Much vibration is caused by the oscillating motion of a single piston. By adding another cylinder (the equivalent of the second circuit in the Tesla motor) a balance is achieved as they reached the top and bottom of their stroke in a synchronous pattern.

Tesla's new conception was not just a new type of motor, however, but an entirely new system of power delivery. In his AEII lecture, and the patents that followed shortly, he established the electrical system that remains largely unchanged to this day.

Within months, Tesla had become a scientific celebrity, much as Edison had a decade before. Unlike the current image of the scientist as emotionless bureaucrat or white-coated automaton, Tesla was seen as a Nietzschean *Übermensch*. John O'Neill, in the first biography of Tesla, repeatedly uses the term Superman to describe him. Like the *Übermensch*, Tesla was a man who sought to overcome human limitations, who sacrificed all for his goal, who was utterly committed to the struggle. In O'Neill's view, Tesla was a man who transcended nature, recreating the human relationship with the world, not a mere technician who created faster and easier ways of performing ordinary tasks.

Financially well-off for the time being, and famous for the ground-breaking work he was doing, Tesla became part of New York's high society. He dined in the best restaurants, spent his evenings with the rich, powerful, and famous. Bankers and journalists, financiers and editors, even Mark Twain and J.P. Morgan, were known to come by his laboratory for demonstrations after elaborate dinners at the Waldorf.

Invitations to his feasts were highly prized, and being asked to the laboratory was an even greater honor. Like Edison, Tesla enjoyed showing off his latest inventions and discoveries. And like Edison, he was now depicted in the popular media as a more-than-human presence. One engraving from the period, published in *The World*, shows him standing tall, lordly, in a fine suit and with a smug expression on his face. Electrical power emanates from his entire body. The caption reads: "The inventor in the effulgent glory of myriad tongues of electrical flame after he has saturated himself with electricity." Even more so than his rival, Tesla put on shows that mingled wonder and danger, amazement and fear.

Bizarre, arcane equipment crowded the lab, making a suitably mysterious background. With his flair for the dramatic, Tesla would have the lights extinguished and as his disciple and first biographer described it:

With invisible fingers set objects whirling, caused globes and tubes of various shapes to glow resplendently in unfamiliar colors as if a section of a distant sun were suddenly transplanted into the darkened room, and crackling of fire and hissing of sheets of flame to issue from monster coils to the accompaniment of sulfurous fumes of ozone produced by the electrical discharges that suggested this magician's chamber was connected directly with the seething vaults of hell.

Nor was this illusion dispelled when Tesla would permit hundreds of thousands of volts of electricity to pass through his body and light a lamp or melt a wire which he held.[8]

More so than anyone else at this time, Tesla explored the shadowy borderland where electrical current and the human body met. Making himself a living conduit, an organic filament, he passed enormous amounts of energy through his flesh. He used his body (or his "meat machine" as he called it) like the transformers he had invented, elevating himself to a sublime state. As the high frequency power passed through him, it became the agent of personal conversion.

With arms outstretched, hundreds of thousands of volts coursing through him, Tesla was the electric avatar of his age, self-crucified, transfigured by his own genius, will, passion.

He understood that the frequency of the alternation was a fundamental difference between electrical current

---

8 O'Neill, 95-96.

and light waves. Electricity could easily kill a person and yet light striking a person couldn't. He surmised that somewhere between the frequency of AC (60 cycles) and that of visible light (billions of times per second) there was a zone of safety.

He divided the damage done by electric shock into two categories. First was the destruction of living tissue by the increase of temperature, which varied in proportion to the change in the amperage of the current. The second was the sensation of extreme pain, which was tied to the number of alternations in the current, each cycle creating a single pain stimulus.

Just as the ear is unable to detect vibrations much faster than 15,000 per second, so Tesla thought, the nerves that send the pain signal couldn't transmit a response if the stimulus had a frequency of 700 hertz (cycles per second) or more. He'd built dynamos that produced current up to 20,000 hertz and tested them by first sending current of this frequency across his finger. And though he felt no pain, he knew that the amperage (the amount of current) was still too great for safety.

With his newly designed air-core transformers, he could increase the voltage by a factor of 10,000 and bring the amperage well within the range that wouldn't harm living tissue. To test his theory, he first passed current through his finger, then his arm, then hand to hand across his body and finally from foot to head. And though the amount of energy carried by the current was enormous, he was in no way harmed. Using this principle, he was able to perform spectacular demonstrations in his lab:

exploding discs made of lead, melting metal rods, lighting both incandescent and vacuum tube lights after passing extremely high frequency current through his body.

He put a related technology to more practical use in lighting his laboratory. By the use of a high-voltage coil, he was able to transmit power without wires. To demonstrate this, he had his laboratory blacked out completely one night and stood in the middle of the room holding two long glass tubes. When his assistant turned the power on to the coil, instantly a powerful blue-white light flooded the room and there stood Tesla waving what he called "swords of fire" as though battling some unearthly, unseen enemy.

Not long afterward, Tesla adapted this technology for the illumination of his entire laboratory. A coil at the ceiling was always energized, so that Tesla and his men could carry a light tube wherever they went in the room and it would produce illumination without any connection to wiring.

Tesla delighted in these displays of wizardry. And a photograph of him, which accompanied his *Century Magazine* article "The Problem of Increasing Human Energy," extended his near-mythic reputation beyond the small group of people whom he invited to his lab. In the picture, often reproduced even to this day as authentic, Tesla used trick photography to create an image of himself as the supreme electrical master. He sits in his laboratory while a spectacular display of lightning tears through the air from a huge central induction coil. In actuality, the time exposure ran close to two hours, producing a much more dramatic display than a shot of a single discharge. And though he appears to sit calmly at the heart of the electrical cataclysm, he had the image of himself superimposed there by double exposure.

Like the "five day vigil" photo of Edison, this picture was also created for publicity. Its main purpose was to raise more capital, and it succeeded. Shortly thereafter J.P. Morgan invested $150,000 in Tesla's work. Whereas Edison stares back at the camera as though in weary defiance, grim and determined, Tesla relaxes, nonchalant at the center of the electrical explosion. Both pictures have been endlessly reproduced, supposedly capturing the scientists in their true element. Both are based on the truth, but stretch that truth to make their subjects bigger than life. Edison promoted himself as the paragon of hard work, warlike will in the face of nature's intransigence. Tesla was much more at ease, not really at war with nature, but by sacrificing himself, at one with it. It was as though he merely needed to align himself properly with nature, put himself in a near-mystical state, and he could glean from her all the knowledge that he desired. Tesla once said of Edison that if he "had a needle to find in haystack, he would proceed at once with the diligence of a bee to examine straw after straw until he found the object of his search. I was a sorry witness of such doings, knowing that a little theory and calculation would have saved him ninety percent of his labor."[9]

For Tesla, theory and practice were indistinguishable; to know was to do. Thought and physical action

---

9   Margaret Cheney, *Tesla: Man Out Of Time*, 32.

were not separate events but reflections of each other. So his displays of monstrous electrical power, his amazing stunts and seemingly life-threatening experiments were just as "true" or as "real" for him as the very practical uses to which he put alternating current.

If the populace fed on images of the Faustian hero, businessmen and financiers were paying close attention to the commercial applications of his work.

George Westinghouse, who became a millionaire inventing and marketing the railroad air brake, was only one of many manufacturers who saw the advantages and huge financial gain to be achieved by implementing Tesla's AC system. At first Westinghouse's forays into the electrical field were tentative, building apparatus to be used in incandescent lighting. But by 1885 he was very interested in the uses of AC technology. In the spring of that year, he read descriptions of the International Inventions Exhibition in South Kensington, England. The inventors Lucien Gaulard and John Dixon Gibbs displayed there a system that included transformers, allowing for greater control and efficiency in lighting. Westinghouse was fascinated and arranged to have some of the Gibbs-Gaulard transformers shipped to the United States for inspection by his engineers. During the tryouts it was found that the transformers weren't suitable, without modification, for Westinghouse's purposes. But convinced there was great value in the new technology, he paid Gibbs and Gaulard $50,000 for the patent rights and had one of his chief engineers, William Stanley, set up an experimental lab in Great Barrington, Massachusetts. Here, Stanley made significant improvements on the transformers and in the spring of 1886 used them to light the town. Alternating current had been employed as early as 1878 (by Elihu Thompson) for arc lighting, but the Gibson-Gaulard system was a great advance and the lighting of Great Barrington was a milestone for the industry.

One of Westinghouse's unused factories in Pittsburgh was refitted for the manufacture of electrical apparatus and within two years he had over 3,000 employees there. The firm was reorganized and the Westinghouse Electric Company was born.

Work went on at a frantic pace. A few months after starting operations, Westinghouse's new concern was able to demonstrate its products. Four hundred electric lamps were supplied with alternating current from a dynamo four miles away. The lamps burned continuously for two weeks, and every day Westinghouse came out to check on the progress of the new system. Once he was satisfied, the dynamo was moved to Buffalo and set up for use in late November of 1886.

Orders for the new technology came in quickly. Greensburg, Pennsylvania—about twenty miles from Pittsburgh—was the first town to have a complete AC system installed for municipal use. With the invention of the Shallenberger meter, to accurately assess the amount of power used by a consumer, Westinghouse was one step closer to making AC profitable.

All that was needed now was a usable AC motor, and here Tesla re-enters the story.

Westinghouse read with great interest of Tesla's lecture to the AEII and a month later wrote a brief note requesting a demonstration. The two men met in Tesla's New York laboratory. Like the first encounter between Edison and Tesla, this meeting was notable for its contrasts. Westinghouse was 10 years older than Tesla, a millionaire business magnate. Powerfully built, with a walrus mustache and the bearing of a man used to success, he stood in stark contrast to the frail, otherworldly man of science. Unlike the meeting in Edison's lab, however, here agreement was quickly reached.

Tesla eagerly demonstrated the AC apparatus. Westinghouse followed him from machine to machine, inspecting each carefully, at times crawling on his hands and knees to get a better look. He quickly understood the brilliance and profitability of Tesla's work and made an offer to purchase the patent rights. The exact amount of the offer has never been established. More than one source claims that Westinghouse told Tesla he would pay $1 million for the rights to manufacture and sell the AC system, and that Tesla accepted with the stipulation that he receive an additional royalty of $1.00 for every horsepower of electricity sold.

The facts of the matter will most likely never be settled, but there is strong evidence that the deal actually entailed $60,000 in cash and Westinghouse stock and a further $2.50 per horsepower in royalties.[10] Given that dozens of dynamos were soon to be built and hundreds more in the ensuing years, the amount due to Tesla

---

10   Cheney, 40.

would have been—if the agreement had been honored—enormous. Just the powerhouses at Niagara Falls, all of them equipped with Tesla machinery, would have netted him millions of dollars.

Whatever the dollar amount, it is certain that a deal was struck. And though Tesla conceived of the polyphase system as a unified invention, it was seen by the United States Patent Office as dozens of individual inventions. In total, Westinghouse bought from Tesla forty patents in 1888.

In addition, Tesla agreed to move to Pittsburgh to work as a consultant to Westinghouse Electric, for $2,000 a month. Though the sizable income was most welcome to a man who had been digging ditches only two years before, it was a hard choice for Tesla to make, demanding that he give up the life he'd been living in New York.

Tesla moved to Pittsburgh expecting to have all the kinks worked out of the system in a year. He quickly ran into technical frustrations and financial constraints. The Westinghouse people had designed all their equipment to work with 133 cycle current and Tesla was adamant that the machinery should run on 60 cycles (which in the end did win out, to become the standard in the United States).

The Westinghouse engineers felt a great urgency to have the AC system perfected and in production quickly; their profit margin was dependent on the speed with which they could get into the marketplace.

Disgusted by what he perceived as small minds,

Tesla quit after spending less than a year in Pittsburgh. Westinghouse tried to lure him back with an offer of more money, but Tesla—now quite wealthy, and frustrated that he'd wasted so much time on trivia and bickering—rejected the offer.

Losing Tesla was a serious blow, but development continued at a rapid pace and soon manufacturing had begun on a variety of motors and dynamos. Westinghouse's goal—nothing less than wiring the entire nation with his AC technology—required not only theoretical expertise, but also huge amounts of capital. The Westinghouse Electric Company expanded quickly, perhaps too quickly, and soon found itself in a precarious financial position.

To make matters worse, two of Westinghouse's main competitors, Thompson-Houston Company and Edison General Electric, merged to form General Electric. With this industrial giant poised for a counterattack, Westinghouse's position was even more tenuous. His financial backers demanded that the company be reorganized to save itself from bankruptcy. A merger was set up to bring together the Westinghouse Electric Company with the U. S. Electric Company and Consolidated Electric. However, before the merger could be consummated, Westinghouse's advisors demanded that he change the legal arrangement with Tesla. They insisted that with the threat of huge royalties looming over the new company, it could never build up the momentum needed to compete effectively with Edison.

It's likely that Westinghouse resisted this proposal. He too was an inventor, and one who'd become very wealthy from his inventions. The patent agreement he'd made with Tesla was fair and legally binding. To ask him to rescind it was both an embarrassment and an affront to Westinghouse's principles. But his backers would not budge, claiming that the money Tesla had been paid was more than enough.

Westinghouse had little hope that Tesla could be persuaded to give up such a large amount of money. He knew that Tesla was still angry and resentful over the treatment he'd received from the engineers in Pittsburgh. Westinghouse's hope was that if he offered an executive position in the newly organized company, Tesla might be persuaded to give up his rights.

The two met where they had met the year before—at Tesla's lab on South Fifth Avenue. Westinghouse explained the situation in a straightforward manner and made his proposal. Tesla asked what would happen if he refused to relinquish his rights. He was told that the Westinghouse company would probably collapse and that the likelihood of the AC system being implemented would be greatly reduced. Edison already hated Westinghouse for AC incursions into DC territory; he would do anything to make sure that DC remained the dominant current.

Most biographies of George Westinghouse omit this event entirely. And in the Tesla biographical literature the meeting of the two men is treated in grand, exaggerated terms, playing up the risk and sacrifice, the friendship and admiration, between the two men.

What actually occurred between them is not known,

George Westinghouse.
(National Portrait Gallery, Smithsonian Institution.)

but the result was that Tesla gave up his rights to the royalties, allowing the reorganization to go ahead as planned. According to the Westinghouse Electric Company annual report of 1897, Tesla was paid $216,000 for outright purchase of patents, freeing Westinghouse from the constraint of any royalty payment.[11]

Though by destroying the contract, Tesla relinquished claim to millions of dollars, he apparently felt no ill will toward George Westinghouse. He trusted him, believed him when he said the change in the contract was necessary. Fifty years later, in a testimonial to the industrialist, Tesla wrote, "George Westinghouse was in my opinion, the only man on this globe who could take my alternating current system under the circumstances then existing and win the battle against prejudice and money power. He was a pioneer of imposing stature; one of the world's true noblemen, of whom Americans may well be proud and to whom humanity owes an immense debt of gratitude."[12]

Thomas Edison was of a different opinion. Calling Westinghouse a "shyster,"[13] he engaged in a war of words as Westinghouse moved into territory previously controlled by Edison. Salvos of derision, claim and counter-claim, were fired back and forth, each touting his product as less expensive and more reliable. Westinghouse had wasted no time in introducing his lighting system into the market, and following Edison's example of ten years

---

11 Cheney, 49.
12 Cited in Cheney, 49 n 16.
13 Andre Millard, *Edison and the Business of Innovation*, 348.

before, priced his product well below cost in order to capture contracts. Within the first two years of availability, 130 towns and cities had Westinghouse AC systems installed. Cheaper and more effective, the Westinghouse system lured some municipalities away from the Edison camp. The pressure on Edison was intense. His agents demanded a more competitive system; some deserted to join Westinghouse. "We could get at least 2,000 or 3,000 lights in," Edison's chief engineer in New Orleans wrote, "at very profitable rates in a nice residential district from one to two miles away if only we had an alternating or continuous current system to support our regular three-wire system. It is idle to scoff at the Westinghouse people—they are hard and persistent workers."[14]

But even before this rivalry developed into full-scale economic warfare, other conflicts were brewing. Tesla's patents—now owned by Westinghouse—were pirated by hundreds of manufacturers and legal action was taken against them. Westinghouse was aggressive, some would say brutal, in his pursuit of these poachers. He won many court battles against infringement, but often the hatred of the losers was directed at Tesla. The so-called Tesla patents were thought of as monopolistic by many engineers, giving Westinghouse a stranglehold on the inventive process.

At the same time Tesla was being attacked on another front. Galileo Ferraris of the University of Turin claimed

---

14   William S. Andrews, letter to J.H. Vail, 12 May 1887. J.M. Vail Papers, Edison Archives, Edison National Historic Site, West Orange, New Jersey.

that he had developed a revolving magnetic field prior to Tesla. Though he had apparently done some work on the principle in 1885, he deemed the idea not practical until Tesla's work was introduced to the engineering world. The *Electrician*, a magazine published in London, made the claim that Ferraris was the true inventor of the rotating magnetic field and that Tesla's inventions were merely an offshoot of Ferraris's work.

While the overwhelming evidence points to Tesla's preeminence in the field, as he had conceived the idea in 1882 and successfully constructed an AC polyphase motor in 1883, the Edison camp took this as an opportunity to attack Tesla, hoping to discredit him as a thief and a liar. The furor over this disagreement died down fairly quickly and today Ferraris is little more than a footnote. But it was a clear indication of the conflict yet to come.

Westinghouse had built up a solid organization and was making serious headway against DC. With millions of dollars at stake, he was quite willing to put all his efforts into dethroning Edison. Always combative, but now with his principles and his sense of identity threatened, Edison was just as eager to lead his followers into battle.

THE ELECTRICAL ENGINEER.
A WEEKLY JOURNAL OF ELECTRICAL ENGINEERING
AUGUST 1888

MR. BROWN AND THE DOG—A BALLAD

(Suggested by the proceedings
at Columbia College, 30 July, 1888)

Then Brown he read a paper, and he demonstrated there,
    That alternating arguments were exceedingly unfair.
No company employed him, and his motives he felt sure,
    Were thoroughly unbiased, philanthropic and most pure.
He had read with deepest feeling that lives were sacrificed,
    And in order to prevent this, new rules he had devised.
At once, as he expected, there arose a fearful din;
    And the alternating advocates hurled weighty papers in.
He took refuge in Virginia, until the skies were clear,
    To prove that he was honest, his "plant" was all in view,
And he called from out the audience three men all good and true.
    One man was set to measure ohms, another watching the volts,
The third to watch the other two, and carefully take notes.

    The dog stood in the latticed box,
        The wires around him led;
    He knew not that electric shocks
        So soon would strike him dead.
    "One hundred volts we'll give him now,"
        The dog did not despair;
    Brown wiped the sweat from off his brow;
        The dog turned not a hair.
    "Two hundred more," then Brown, he cried,
        The dog did not complain:
    He knew not that his friends had died—
        That he must die in vain.
    At last there came a deadly bolt;
        The dog, O where was he?
    Three hundred alternating volts
        Had burst his viscerae.

EPITAPH
Every dog must have his day;
    He had his and passed away.
He never drained life's bitter cup,
    Death took him when he was a pup.

Chapter Five
THE WAR OF THE CURRENTS

On June 4, 1888, New York's Governor Hill signed the electrocution bill into law, to go into effect January 1, 1889. On June 5, 1888, a curious letter appeared in the *New York Evening Post*. Written by Harold P. Brown, "Death in the Wires" describes in great detail the dangers "inherent" in alternating current. Evoking the memory of a little boy who'd touched a downed telegraph wire on East Broadway and was killed by the current, Brown warns his readers that alternating current, if allowed to triumph over direct current, would pose a grave threat to public safety.

This letter marks the entrance of a figure, now largely forgotten, who between 1888 and 1890 came to stand center stage in the drama of the AC-DC battle. Though he had no formal education in electrical science, though he was not a leading engineer, businessman or inventor, Harold Brown would play a crucial role in the most public chapter in the struggle between Westinghouse and Edison.

Brown's letter, which two newspapers refused to publish, is noteworthy both because of its timing and because it was a clear and substantive attack on the Westinghouse system. Published in a popular periodical, and aimed at the general reader, the letter paints a picture of alternating current as a "damnable threat," an example of the Westinghouse "combine's" supposed greed and utter unconcern for the safety of the public. In the letter, Brown calls for the New York Board of Electrical Control to forbid the "fatal" high voltage current as Chicago had done. Public safety and public interest, Brown claimed, were all that induced him to write. He knew that his letter would stir heated emotions, he admitted, but he felt the altruistic necessity of telling the truth. Suggesting regulations that would have banned high-voltage AC, "Death in the Wires" was read before the Board of Electrical Control three days after it was published in the *Post*. It then went on to local electric companies for their reaction. Also, the board invited anyone desiring to answer Brown's assertions to attend its next meeting at Wallack's Theater, July 16, 1888.

Brown had been hired at age nineteen by Western Electric in Chicago and was soon placed in charge of Edison's inventions, including the "electric pen." In 1879, he went to work for the Brush Electric Company, and stayed with them for five years. Brush had pioneered the development of arc lighting in the United States and business expanded rapidly during the time Brown spent with the company. In 1884, Brown left Brush to form his own electric company. His letterhead at this time includes the following description of services he could provide: Designer of Apparatus for Special Purposes, Contractor for Arc and Incandescent Electric Lights and Steam Power, City Street Lights Tested and Compared with Contract Requirements, Complete Plants Erected for City Lighting. In the last half of the 1880s Brown was granted a number of patents for inventions combining incandescent and arc lighting on the same high voltage circuit.

Aside from working with the Edison patents ten years before, from 1876 to 1879, Brown had no connection with Edison or his organization. F. S. Hastings, the secretary of the Edison Electric Light Company introduced Brown to his employer, who was favorably impressed with the young engineer. At this time Edison allowed many individuals to use his facilities for private experimentation. Brown requested to work at the West Orange lab, using equipment not available elsewhere. Edison agreed, and instructed Arthur Kennelly to work with Brown. A colleague of Brown's, Dr. Fredrick Petersen, a nerve specialist at New York's Vanderbilt Hospital, was also welcomed there. It's likely that Brown's point of entry with Edison was his claim to be working on improving DC's safety. Edison had a long-standing dread of accidental deaths associated with "his" current, and any possibility of reducing this risk would certainly have appealed to him. Also at this time, Edison and his West Orange staff were hard at work on perfecting the phonograph (hoping to have it done by Christmas of 1888) and any diversion was probably welcome.

On July 16, the Board of Electrical Control met at Wallack's Theater. At that gathering, various letters and oral testimony were presented. A vice president of the Westinghouse Company and the secretary of the American Institute of Electrical Engineers were among those whose opinions of Brown's letter were made known. More than one engineer treated Brown's ideas as absurd, ill-conceived and based either on gross ignorance or guileful manipulation of the facts. Just as he accused Westinghouse of twisting the facts to suit his needs, so the proponents of AC asserted that Brown was misleading the public for financial gain. The Board was informed that Brown was promoting an apparatus that converted dangerous high voltage currents for incandescent lighting and that the success of alternating current was seriously eroding its value in the marketplace.

However, mere economic gain and loss are insufficient to explain Brown's dogmatic hatred of the Tesla/Westinghouse system. Why Harold Brown became entangled in the AC-DC fight may never be determined definitively. Association with Edison, the most famous inventor of the day, certainly had its allure. Brown went from obscurity to notoriety in a few months. He was called on by various political bodies as an "expert," "authority," "well-known scientist." Perhaps the allure of fame combined with the call of money was enough. But the virulence of his crusade against AC indicates something deeper was at work. Perhaps it was the case of a mediocre man thrust into the limelight and then working frantically to keep his mediocrity and dearth of ideas hidden. To admit he was wrong, to admit he was merely motivated by greed and self-aggrandizement when he'd touted himself as a benefactor of humankind would have been too painful. Whatever the reason was, Brown's next step was to begin a series of experiments that would "prove" his assertions were correct.

Brown sent out invitations and on July 30, 1888, a group of scientists, reporters, and members of the Electrical Control Board gathered in Professor Chandler's lecture hall at the Columbia School of Mines in New York. The July air was heavy with heat and humidity. Shining lacquered paintings stared down from the polished paneled walls. A chalkboard and lectern stood to one side. Stacks of equipment and metal cages crowded the lecture stage.

Brown took the podium and began by reading a paper defending his ideas. "I've not become involved in this controversy because of any connection to financial or commercial interests. I am here only because of my sense of right. There are three classes of current in commercial use: the continuous, the intermittent, and the alternating. Investigation has shown that the first and second could by proper safeguards be made harmless to the general public, while the third is by its very nature hopelessly deadly."[1]

He went on to describe earlier experiments proving, he claimed, that living creatures could stand shocks from continuous or direct current far better than from

---

[1] Dialogue from the July 30, 1888, experiment is paraphrased. See *New York Times*, 31 July 1888.

alternating. "I have applied a current of 1,410 volts—continuous current—to a dog without fatal result. And I have repeatedly sent to eternity dogs with as little as five hundred volts alternating current. Those advocates of the alternating system who claim to have withstood shocks of 1,000 volts without injury must have been wearing lightning rods. Surviving such a shock is impossible, and those who state otherwise are compelled either by ignorance or guileful commercial interest."

Besides Arthur Kennelly, on the stage with Brown that day was Dr. Petersen. A physician who practiced for four years in Buffalo and then three years in a lunatic asylum in Poughkeepsie, Petersen was one of the country's foremost experts in "medical electricity." He had studied mental and nervous disease in Vienna, Strasbourg, and Leipzig and at the time of the Columbia lecture had been using electricity for about four years in his practice in New York. During this period he was applying current on average thirty times a day in the hospitals and dispensaries he was affiliated with.

"To demonstrate the veracity of my beliefs," Brown went on, "I've asked you gentlemen here today to witness the experimental application of electricity to a number of brutes."

Brown's assistants led onto the lecture stage a large dog. At seventy-six pounds, the Newfoundland was a strong and healthy animal. Brown had him put into one of the wire cages, then muzzled and secured with belts. Immobile now, probed and manhandled by the scientists on the stage, the dog whined and growled anxiously. By means of a device called a Wheatstone bridge, the dog's resistance was measured at 15,300 ohms.

"We will first apply the continuous current at a pressure of three hundred volts," Brown announced. He made sure the wires leading to the animal were secure, then nodded to Kennelly. The switch was thrown and the animal yelped and thrashed helplessly in the cage. A relay attached to the apparatus shut the current off after a short shock. "Observe, gentlemen, that our subject though discomfited, is still quite healthy." The voltage was increased to 400 and again the current blasted into the dog's body. His complaints and twitchings were much more noticeable now. Drool spattered the floor of the cage, his eyes flashed, growls and a sudden gurgling sound—as the air was squeezed from his lungs—reached the audience. Some of the observers murmured their protest. One got up to leave. But before he reached the door, Brown's voice called out, "We increase now to 700 volts." The current ran; the dog's struggle was desperate now. Putting all his strength into his attempts to free himself, the dog broke the muzzle and began gnawing at the leather straps that held him in place. He was subdued though, muzzled again and his resistance rechecked. It had fallen in only a few minutes to 2,500 ohms, one sixth its earlier level.

"Finally, we increase the current to 1,000 volts."

Groans came from the audience. A man waved his cane and shouted, "No!" but the current was applied again and the dog this time shook and thrashed pathetically. The smell of singed hair filled the air. The dog's guttural noises could be heard above the hiss and hum of the machinery.

When the current was cut, Brown took the podium again and said, "He will have less trouble when we try the alternating current. As these gentlemen say, we will make him feel better."

With a pressure of 330 volts, AC, the dog was quickly put to death. He was untied and his muzzle removed. A smell of meat, burnt hair, and ozone lingered long after the body was removed from the lecture hall. Brown announced he would use alternating current first on the next subject. The dog was led out but before he could be tied and muzzled, a man rose from the audience and demanded that the experiment stop. He identified himself as Agent Hankinson of the SPCA, showed his badge, and told Brown that no more dogs would be killed that day. Brown spoke briefly with Hankinson and agreed to this demand. The dog was led back to the basement and Brown returned to the lectern.

"The experiment is a sham!" a man shouted from the front row. He stood and glared at Brown. "Your conclusion that continuous current is less dangerous than alternating is based on the fact that you applied the continuous current first. The beast was almost dead when the alternating was applied. If you'd preceded the other way—"

"That, sir, was my intention," Brown said. "If the experiment had been allowed to continue."

Another advocate of AC spoke up, pointing out that there had been no relay on the AC system, so the current hitting the dog had gone on a great deal longer than the sudden bursts of DC administered first. "I have a proposal, Brown. If the gentleman from the protective agency will not allow you to proceed further with the animals, why not put yourself in the beast's place? If direct current is safe, you should have no concern."

Brown pretended not to hear the question and announced that the demonstration was ended. "The only places," he said, "where alternating current ought to be used are the dog pound, the slaughterhouse and the state prison."

A few days later, Brown was back at the Columbia School of Mines with a similar set of experiments. On August 3, he killed three dogs using alternating current of a pressure less than 400 volts. Reminding the audience that he had already demonstrated that a dog could withstand a voltage of 1,000 DC, he felt confident that he'd proven beyond a doubt that alternating was far more dangerous than direct current.

Brown wrote to Arthur Kennelly the day after the second set of experiments: "It is certain that yesterday's work will get a law passed by the legislature in the fall, limiting the voltage of alternating current to 300 volts." He also told Kennelly that he'd "lost 12 pounds over this struggle and am all worn out, but am going to the mountains today to rest."[2] However, the next two years would allow Brown little time for relaxation.

Throughout the remainder of 1888, Edison's West

---

2   Harold Brown, letter to Arthur Kennelly, 4 Aug. 1888, Edison Archives.

Orange laboratory was the site of experiments performed by Brown and Edison's staff. In an atmosphere that might be described as a cross between a circus sideshow and an abattoir, experimenters—or "muckers" as Edison liked to call them—went about their work often with little supervision from upper level staff. Employing over a hundred men, the West Orange lab was a buzzing hive of scientists, amateurs, enthusiasts, teenaged aficionados, cranks and con men. Explosions rocked the lab from time to time; men worked on projects with no specific goal or hope of success. At one point, several of the muckers were instructed to produce extremely toxic chlorine gas. Chloroform was used as an antidote, and workers could be found lying about the lab, sniffing the gas like addicts in an opium den.

It was in this atmosphere that Brown's grotesque experiments went on through the fall of 1888. Though working with his characteristic tenacity on the phonograph, Edison himself joined in, and at times many of the most trusted members of his staff were also participants in the grim proceedings. How many animals were killed in the electric slaughterhouse cannot be determined precisely. Certainly, though, dozens and probably hundreds were hooked up to the alternating current dynamos and "sacrificed on the altar of science," as the newspapers liked to call it. Edison wrote to the SPCA, requesting that they provide him with dogs for his experiments. The request was refused. Then he offered a bounty of $.25 for every cat and dog brought to the facility. Such was the demand that there were times when the local boys would be lined up outside the lab with their catches. Once the strays were all gathered up, no pets were safe. With a price on their heads, many were spirited off and never again seen by their owners. It was found eventually, though, that cats had a peculiarly high resistance to electricity and were dropped in favor of dogs.

The dynamo room at the West Orange lab was the main site of the experiments. Crowded with generators, transformers, relays, and cables, the room had a close, near-claustrophobic feel to it. The traditional machine shop smell—grease and sweat, singed metal and overheated drive belts—was soon tinctured with the scent of cooked flesh and burnt hair. One end of the room was taken up almost entirely by two broad arched windows. The walls of wooden paneling were polished to a bright sheen. And because much of the experimentation took place at night (both to keep the curious public from disturbing the work and so that the equipment could be used by day for more traditional purposes), the walls and windows mirrored the flashes and sprays of man-made lightning. The room echoed with howls and squealing of animals attached motionless to the dynamos. The room had an air of both secrecy and boyish sadism. At one point, Charles Batchelor, attempting to hold a puppy in a makeshift electric chair, received a powerful shock himself, and "had the awful memory of body and soul being wrenched asunder... the sensation of an immense file being thrust through the quivering fibers of his body."[3]

Dr. Petersen was often present after the electrocutions,

---

3   Josephson, 347 n 11.

dissecting the subjects and examining nerve and blood cells under the microscope. The corpses were taken away to be incinerated. In order to best shape the results of the experiments, Edison forbade his staff to make any public comments on the killings. Though Edison Electric Light Company paid all the costs, they were billed to secretary Hastings. The enormous scale of the work that autumn made the experimenter's motives seem dubious. Edison did at times invite news reporters to witness the experiments, yet he knew the slaughterhouse atmosphere might taint all electrical current in the public mind. He and Brown went to great lengths to make certain the distinction between "killer" AC and his benevolent DC was clear to reporters and politicians who came out to view the bloody proceedings.

Brown's work continued to be criticized by proponents of AC, especially the fact that the animals he killed with AC weighed much less than the average human being. To answer these critics, Brown brought the experiments to a spectacular climax on December 5, 1888. Aware that the date when the electrical execution law would go into effect was less than a month away, Brown made a demonstration before Elbridge Gerry and the committee appointed by the Medico-Legal Society to investigate the problem of how best to replace the gallows with the electric chair. On the afternoon of December 5, assisted by Arthur Kennelly, Brown electrocuted two calves and a horse before a gathering of scientists, doctors, and government officials. The first subject was a 124-pound calf. The animal was cut on the spine and a sponge-covered plate moistened with zinc sulfate solution was fastened in place. And alternating current of 700 volts was applied steadily for thirty seconds and the animal collapsed, dead. One of the metal electrodes had made contact with the animal's forehead. Other than the burning at that spot, no indication of external injury was found. The calf was immediately dissected by Doctors Ingram and Bleyer. The lungs, brain, and heart were found to be in a "normal condition" and the meat pronounced fit for human consumption. A second calf weighing 145 pounds was next killed by the alternating current.

"To settle entirely and conclusively the question of weight," Brown declared to the audience, "our last subject is a horse weighing 1,230 pounds."[4] The electrodes were attached to the forelegs so that the current would pass through the animal's trunk. When the current was turned on, the horse snorted, whinnied and shook, but the 700 volts was sufficient and like the calves, the horse crumpled, dead, before the satisfied spectators.

With the three dead animals cut to pieces, the air rich with the smell of blood and ozone, machine grease and cooked flesh, Edison's lab that day was less a slaughterhouse than a chamber in some temple where propitiatory offerings were given to bloodthirsty gods. Though the patina of rationalism covered the work, though a nominally scientific desire for knowledge informed the proceedings that day, it's not difficult to see this event

---

4   Dialogue from the 5 Dec. 1888 experiment is paraphrased. See *New York Times*, 6 Dec. 1888.

as a brutal hecatomb in honor of an unnamed (perhaps unnameable) deity. The altar might have been built in a machine shop and the agent of death carefully calibrated and controlled, but the actual event stripped of its technological trappings could be seen as a sacrifice. The doctors in their blood-spattered aprons, the laborers unfastening the dead beasts and dragging them to their places on the electric altar, the observers—priests whose duty it was to maintain social order, justice, and peace—might have been performing their tasks thousands of years before in a torch-lit cave. The fact that the doctors declared the meat fit for consumption adds to the perverse sense of sacrifice. The blood of the animals in Edison's lab, like that of the criminals who waited on death row, may have served a deep, sacrificial purpose.

Earlier that year the New York legislature had charged the Medico-Legal Society with working out the technical details of electrocution (left ill-defined in the law signed by Governor Hill). At the demonstration on December 5 the society's special committee was given a glimpse of electrocution's promise. By this point, however, there were few questions left unanswered. It was certainly not a coincidence that Frederick Petersen was the chairman of the committee, nor that Edison—not Westinghouse or Tesla—was consulted at great length regarding electrocution.

Brown and Petersen collaborated on the report for the Medico-Legal Society and it was no surprise that when the society met on December 12 for its annual banquet it approved the report unanimously. The report spells out exactly which type of current—AC of no less than 300 cycles per second—should be used for electrocution. It asserted that death by AC would be instantaneous and painless, while that caused by DC would be accompanied by "howling and struggling."[5] The report also recommends which posture the prisoner should be in when the current is applied. Standing was deemed unacceptable, lying down better; sitting in a chair was the preferred position. Unlike the final version of the electric chair, the one described in the report would have applied one electrode to the prisoner's neck and the other only a few inches away at the top of the head. Voltage from 1,000 to 1,500 was recommended, administered from 15 to 30 seconds. With the new law going into effect on January 1, Brown and the DC interests had succeeded in getting all in order with only a few weeks to spare. The adoption of the report and the publicity that accompanied it achieved for Brown and Edison one of their most important goals. AC was—at least in the eyes of the state of New York—the killer current.

The same day that the Medico-Legal Society's meeting was reported, a letter from George Westinghouse appeared in the *New York Times*, challenging the validity of the December 5 experiments. Westinghouse's letter accuses Brown of manipulating the experiments and asserts that Brown was in the "pay of the Edison Electric Light Company; that the Edison Company's business can be vitally injured if the alternating current apparatus continues to be as successfully introduced and

---

5   *New York Times*, 13 Dec. 1888.

operated as it has heretofore been; and that the Edison representatives from a business point of view consider themselves justified in resorting to any expedient to prevent the extension of this system." Furthermore, the experiments had little practical application, according to Westinghouse, because Brown had passed the current from brain to spine in the animals—which was very seldom the path it took in accidents—and because he artificially lowered the resistance by cutting into the beasts' flesh and using a salt solution for better flow. Defending the safety of AC, Westinghouse claimed that it was "less dangerous to life, from the fact that the momentary reversal of direction prevents decomposition of tissues, and injury can only result from the general effects of the shock; whereas in continuous current there is not only injury from the latter cause but a positive organic change from chemical decomposition, much more rapid and injurious in its effects."[6]

Westinghouse points out that his company had orders for 48,000 lights in the month of October 1888 alone, whereas Edison—according to its own annual report—had sold only 44,000 for the entire year. Westinghouse then states that he has "no hesitation in charging that the object of these experiments is not in the interest of science or safety, but to endeavor to create in the minds of the public a prejudice against the use of alternating current." Concluding, Westinghouse suggests that the *New York Times* send a disinterested and competent investigator to ascertain the facts in the dispute.

---

6   *New York Times*, 13 Dec. 1888.

Brown's reply was printed five days later. He in turn accuses Westinghouse of fudging and ignoring the facts to suit the needs of his business, and then upping the stakes, suggests that the two of them take part in a duel.

> I therefore challenge Mr. Westinghouse to meet me in the presence of competent electrical experts and take through his body the alternating current while I take through mine a continuous current. The alternating current must not [have a frequency of] less than 300 alternations per second (as recommended by the Medico-Legal Society). We will commence with 100 volts, and will gradually increase the pressure 50 volts at a time, I leading with each increase, until either one or the other has cried enough, and publicly admit his error.[7]

Aside from the questionable scientific validity of such an experiment, the symbolic value of the challenge added some weight to Brown's position. Like an inquisitor demanding that a heretic recant his fallacious belief or face trial by fire, Brown attempted by his challenge to reach higher ground. An issue of faith, a test of moral resolve, the trial would prove which side truly believed and which was driven by greed. When Westinghouse declined the challenge, Brown pointed out that his foe had no qualms about endangering the average person on the street but wouldn't place himself in the same situation. He also mentioned a rumor that Westinghouse—for

---

7   *New York Times*, 18 Dec. 1888.

reasons of safety—had direct, not alternating, current installed in his own home.

During the following year, the war moved into a number of fronts—personal attack, further public demonstration and experiment, lobbying political powers, and publication in leading magazines of the day. Edison drew on his reputation as one of America's most admired citizens to fling contempt and derision on his enemy. Edward Dean Adams, a Boston financier who was both a friend of Westinghouse and a member of the Board of Directors of Edison's company, tried to negotiate a truce between the two men. In response to Adams's suggestion that Edison go with him to Pittsburgh to visit Westinghouse's facility, Edison sent a caustic telegram: "Am very well aware of his resources and plant, and his methods of doing business are lately such that the man has gone crazy over sudden accession of wealth or something unknown to me and is flying a kite that will land him in the mud sooner or later."[8] Some of Edison's men, with a better grasp of the science in question, begged him to give up his vehement opposition to AC. And in fact between 1888 and 1892 much work was being done in the Edison labs on AC, particularly by Arthur Kennelly. Experiments with high-voltage DC were also taking place.

Edison knew that his company was in the best position to take advantage of the AC technology. Still, he would not or could not relent. As Harold Passer points

---

8   T.A. Edison, telegram to E.D. Adams, 2 Feb. 1889, Henry Villard Papers, Edison Archives.

out, "In 1879, Edison was a brave and courageous inventor. In 1889, he was a cautious and conservative defender of the status quo."[9] In 1879, scientists, engineers, and of course the gas companies that he threatened, declared that Edison's ideas were unsafe and impractical, that he was merely following an ignis fatuus. Ten years later, in a book titled *A Warning*, published and distributed by Edison, the believers in AC were accused of chasing the fool's light. George Westinghouse was, "the inventor of the vaunted system of distribution which is today recognized by every thoroughly read electrician as only an ignis fatuus, in following which the Pittsburgh company have sunk at every step deeper into a quagmire of disappointment."[10] *A Warning*, bound in gold, was one of the first attempts through the print medium to affect the outcome of the AC-DC fight. With the title in bold red letters, *A Warning* outlines the dangers "inherent" in alternating current, and includes a list of those supposedly killed by the "executioner's current." The booklet also alleges that Westinghouse and Thompson-Houston—a company also moving rapidly into the AC field—were "patent pirates" of a decidedly low moral character.

Harold Brown published a book called *The Comparative Danger To Life of the Alternating and Continuous Currents*, a compendium of speeches, reports, newspaper articles, and experimental evidence in support of the DC position. No publisher is indicated on the book, but the cost of its printing and distribution

---

9   Harold Passer, *The Electrical Manufacturers, 1875-1900*, 174.
10   Edison Electric Light Company, *A Warning*, 72.

was very likely covered by Edison. Public declarations notwithstanding, Brown was clearly on Edison's payroll by this point and his efforts over the next year and a half were almost entirely shaped by Edison's agenda and money. *Comparative Danger* treats in detail Brown's experiments in electrocuting animals and also includes a death-roll of those purportedly killed by alternating current. Brown also sent letters to the New York newspapers, listing ten victims by name and alleging there were many others.

Westinghouse countered by having his sales agents investigate Brown's allegations. A summary of the findings shows that of the thirty deaths Brown attributes to the Westinghouse system, in twelve cases there was no Westinghouse plants in the city at the time of the deaths, one was a drowning, sixteen were caused by direct current, or no death occurred at all. Only one, according to Westinghouse's research, was directly caused by AC.

Early in 1889, Brown sent a circular letter to the mayors, ranking politicians, insurance agents, and notable business people in every town and city in the United States with a population over 5,000. It begins, "I address you in a matter of LIFE AND DEATH which may personally concern you at any moment." Describing Westinghouse as "criminal," "careless," "both wealthy and unscrupulous," "vicious," Brown attempts through scare tactics what he failed to do through scientific demonstration. Calling on the name of Thomas Edison, claiming the authority of the Board of Health and the Medico-Legal Society, Brown again rails against the "alternating current syndicates" like a righteous crusader whose only motives are truth and higher good. He mentions by name individuals who were "killed and crippled for life," "paralyzed," "completely shattered," then asks in boldface letters "HOW DOES THIS AFFECT YOU?"

He suggests a remedy to "Keep this EXECUTIONER'S CURRENT out of our homes and streets and prevent reckless corporations from saving their money AT THE EXPENSE OF THE LIVES OF THOSE DEAR TO YOU." Simply put, he wanted all towns and cities in America to limit the pressure in electric lines to 300 volts, which would have effectively put Westinghouse out of business. After assuring his readers that he would be glad to assist them "in any way against the encroachments of the executioner's current," he asks them in return to send him information regarding the deaths or injuries from electricity in their area.[11]

Westinghouse returned fire with a booklet, *Safety of the Alternating System of Electrical Distribution*, in October of 1889. Besides explaining that fire hazard was less with AC than DC, because there's less chance for a spark to jump a gap in an open AC switch, he emphasized that alternating transmission, because of the transformers involved, brought much lower voltage (50) into the home than did DC (110).[12]

Elbridge Gerry weighed into the propaganda war with an article published in the *North American Review*

---

11   Brown's circular reprinted in *Electrical Engineer*, Feb. 1889.
12   George Westinghouse, "Safety of the Alternating System of Electrical Distribution," 27.

and titled "Capital Punishment by Electricity." He avoids any technical considerations and argues his case from a legal, moral, and spiritual standpoint. After evoking Mosaic laws that demand the death penalty, describing the horrors of a botched hanging ("The noose tightened, however, and after twenty-four minutes, during which the body writhed and twisted"), and giving a brief overview of the process New York went through to arrive at its electrical death law, he asserts that electrocution may be unusual but will not prove to be cruel. He concludes his brief essay by stating that "to contend that a proper electrolethe cannot be constructed, certain in its effects, when such fatal accidental results from an ordinary arclight dynamo, intended only for illuminating purposes, are proven beyond dispute, is simply to argue an absurdity."[13]

Edison, too, added his voice to the public debate in the same magazine. In "The Danger of Electric Lighting," he evokes his status as America's greatest inventor to further cast doubt and suspicion on Westinghouse. Interestingly, he calls into question the idea of "the expert" and expert testimony, as though his standing were somehow far more elevated and pure than that of mere electrical engineers. Like Brown and Gerry, he couches his arguments in terms of public safety, attempting to raise himself above crude economic squabbling.

In the same issue of the *North American Review*, Brown's article "The New Instrument of Execution"

---

13   Elbridge Gerry, "Capital Punishment by Electricity," *North American Review*, Sept. 1889.

appeared. In it he goes over the same arguments and experimental data he had used in earlier publications. What is notable about the essay, though, is that it gives a glimpse of what the electric chair would look like. He suggests that the prisoner should be "shod in wet felt slippers" and instead of electrodes to apply the current, his hands and feet should be immersed in tubs filled with potash solution. "The deputy-sheriff closes the switch. Respiration and heart-action instantly cease... There is a stiffening of the muscles, which gradually relax after five seconds have passed; but there is no struggle and no sound. The majesty of the law has been vindicated, but no physical pain has been caused."[14]

Brown's propaganda efforts were not limited to the printed page, though. He moved into active political lobbying. Along with Edison, Brown appeared before a committee of the Virginia legislature, which was debating a bill for the "prevention of danger from electrical lighting." Their goal in Virginia, as in similar situations, was to have the government limit voltages to 300 or less, which would eliminate the major advantages of AC. Though their testimony did not sway the Virginia lawmakers, Brown continued to lobby other states. In Columbus, Ohio, he performed another set of animal experiments as a bill was being considered in the Ohio legislature to limit voltages. In Pennsylvania, New Jersey, and other states, Brown took his message and his show on the road.

---

[14] Harold P. Brown, "The New Instrument of Execution," *North American Review*, Nov. 1889.

During this same period, Brown was preoccupied with the actual implementation of New York's electrocution law. In March 1889 Carlos F. MacDonald, superintendent of the State Asylum for Insane Criminals at Auburn, wrote to Brown requesting that they meet so that Brown could present a detailed proposal for his design of the death apparatus. The next week, Brown met in Albany with Austin Lathrop, the state superintendent of prisons. The meeting was successful; the next day Edison received a letter informing him that Brown had been authorized by New York to proceed with the construction of the death machine. In its announcement of the deal, the *New York Times* mentioned that "in addition to a Westinghouse dynamo, each prison will require an 'exciter,' to be used as an auxiliary to the dynamo; a strong oaken chair, in which the convict is to sit and be killed; and electrical shoes."[15]

The one hitch, however, was Lathrop's stipulation that the state pay for the generators only when they'd proved effective. Again going to the source, Brown asked Edison for $8,000 to buy the dynamos. But money was not the only impediment to the successful use of the enemy's technology in the death chair.

Westinghouse predictably refused to sell the generators to Brown, so he was forced to use less than direct means to acquire them. To solve his problem, Brown decided to go through a third party. He discussed the matter with Edison and apparently because of Edison's influence, the Thomson-Houston Company—one of

---

15   *New York Times*, 8 May 1889.

Westinghouse's main competitors in the AC field and already involved in the negotiations that would lead to consolidation with Edison to form General Electric—agreed to act as the intermediary. Using a Boston dealer of used electrical equipment, Thomson-Houston was able to buy three Westinghouse generators. C.A. Coffin, the company's treasurer, made arrangements to reimburse Brown for the purchase of the machinery and also provided an additional $1,000 for expenses related to the electrical tests performed by Professor Louis Duncan at Johns Hopkins University.

Brown was convinced that Westinghouse would go to any expense to foil his plan. So, besides the use of a middleman to hide the identity of the party actually buying the equipment, great precaution was taken to keep the nature of the machinery hidden while it was transported to Auburn. Shipped in waterproofed, unmarked, crates, stored under lock and key, the generators reached the state prison in June of 1889.

Details of Brown's machinations were revealed in August of that year when his office was broken into and forty-five letters stolen. Published that month in the *New York Sun* under the headline "For Shame, Brown—Disgraceful Facts About the Electric Killing Scheme," the letters make it obvious to what degree Brown and Edison would go to make certain AC was used for the death chair. Secrecy, nameless middlemen, subterfuge, exaggeration, and threats are in evidence. As though one brute should be destroyed by another, he mentions that one of the generators, like some crazed animal, had the reputation of being a "man killer." That the letters were authentic is suggested by Brown's strenuous efforts to have the thief punished. He offered a $500 reward for information leading to the conviction of the person responsible and made it known that he would present his case to a New York City grand jury.

But according to the *New York Sun*, it was Brown who should be punished. In an editorial, the *Sun* accused him of venal and corrupt motives, claiming that a conspiracy existed against the Westinghouse company and that Brown was the chief conspirator. The editorial then recommended that New York State have nothing more to do with Brown, as he was unfit—both morally and professionally—to be involved in such sensitive and important matters.

The public debate continued to flicker and flair in the pages of periodicals, and even in the pulpit. Hugh O. Pentecost, a popular speaker, addressed a few hundred people in July of 1889 on the topic of "Murder by Law." While his sermon was centered around an upcoming multiple hanging in New York City's Tombs, he used the opportunity to attack Harold Brown. He referred to the electric chair as "the most diabolical contrivance that the human mind had ever conceived." And regarding Brown himself, Pentecost called him a "lizard-blooded scientific promoter of murder, a creature to be forever loathed."[16]

The *New York Evening Post*, earlier a supporter of electrocution (when it was presumed that the prisoner would be killed by his merely touching an electrified

---

16   *New York Times*, 12 Aug. 1889.

knob), changed its position upon release of pictures of the "terror-giving paraphernalia."[17] with titles such as "Electric Murders," "Electric Wire Slaughter," and "Another Corpse in the Wires," letters to the editors of many papers further aggravated the controversy.

Addressing the convention of the Natural Electric Light Association, Dr. Otto Moses accused Brown of cruel and Machiavellian calculation, and encouraged the members to reject his arguments, as electrocution would only bring shame and doubt on the profession. He later called for the New York State legislature to repeal the electrocution law. And even C. A. Coffin, who'd been instrumental in providing money to buy the AC generators, expressed his doubt whether the voltage produced by the dynamos would be sufficient to cause instant and painless death.

While the furor about the electric chair's acceptability boiled, another debate was taking place over what to name the new process. As the technique had never been used before, there was no standard term for it. And while this might seem a trivial consideration, even on this level the conflict had personal, as well as professional, ramifications.

Elbridge Gerry had suggested the term "electrolethe." *Scientific American*, attempting a more clinical gloss on the process, included in its list of possible terms: "thanelectrize," "electropoenize," "electrostrike," and "fulmenvoltacuss." In response to a *New York Times* editorial suggesting the term "electrocution," readers wrote in

---

17   *New York Evening Post*, 14 May 1889.

Original electric chair, Auburn prison.

The Auburn prison chair, in a later location.
(Both photos courtesy of John Miskell.)

with their own ideas: "electrolysis," "electricide," "electricission," and "electrothanasia." Edison proposed the terms—"ampermort," "dynamort" and "electromort." But the name he liked best was "Westinghouse." Just as the guillotine, Edison reasoned, was named after Dr. Guillotine—so the electric chair should be named after the man whose greed and cruelty made it possible: George Westinghouse. The term, he suggested, could be used as both noun and verb: the prisoner was condemned to the westinghouse where he was westinghoused. The *New York Sun*, following the same line of reasoning argued that the process should be named after its true father, Harold Brown. However, the word "electrocution" (at first in quotation marks) was used in the year of Kemmler's death and was soon generally accepted.

Authorities at Auburn Prison, where Kemmler waited in solitary, acknowledged on June 7, 1889, receipt of the AC generator. Two months later, Brown and his so-called assistant arrived at Auburn to inspect the machinery and supervise construction of the electric chair. The "assistant"—Edwin Davis—was in 1889 an unknown electrician. Harold Brown would sink back into obscurity, but Davis would gain prominence and remain in the public eye for years as the world's first electrocutioner. It would be Edwin Davis who would throw the switch that routed the current into Kemmler's body, and who would perform the same task 239 more times between 1890 and 1914.

A short, spare, wiry man with piercing eyes and a drooping mustache, Davis—though he tried throughout his career to stay out of the public eye—came to take an important place in the mythos of the electric chair. After making improvements on the electrodes, Davis carried the patented devices in a black bag, like a doctor on a house call. He guarded the electrodes jealously, allowing no one too close a view, and refusing to train a successor until very near the end of his career. Wearing a black felt hat and a somber Prince Albert overcoat, Davis could be seen traveling by train throughout New York, and later, New Jersey and Massachusetts, too, after those states adopted the chair. Quiet, reserved, devoted to the bees he kept as a hobby, Davis gained legendary status as the unknowable little man whose job was death.

Brown's intention during this visit to Auburn was to perform experiments in the presence of witnesses, in particular Carlos MacDonald and George Fell. But schedules didn't allow, yet, for the four men to meet. Questioned again by a reporter about the ability of the apparatus to "knock the life out of Kemmler," Brown reeled off the same arguments he'd been making for a year and asserted he had a list of ninety men who'd been killed by electricity, fifteen by Westinghouse dynamos identical to the one already installed in Auburn, Sing Sing, and Clinton Prison in far-north Dannemora.[18]

A special commission was appointed to test the effectiveness of the generators. Doctors Carlos MacDonald and A.B. Blackwell served on the commission, as did Professor Louis Laudy of the Columbia School of Mines. Traveling first to Sing Sing, they inspected the newly

---

18   *New York Times*, 6 Aug. 1889.

installed machinery and, under Professor Laudy's supervision, tested the voltage produced by the dynamo. After some delays due to faulty drive belts, they lit a battery of light bulbs and pronounced the dynamo sufficient for its intended purpose. Sing Sing's death house—holding the electric chair and four death row cells—had recently been built by prison labor. While the commission was watching the dynamo in action, not far away laborers were working on the interior of the small brick building.

Only a few days later, the commission was in Auburn, joined now by Austin Lathrop, and Dr. George Fell, who was there to test his pulmotor resuscitation device on the animals shocked that day. Because of the sensationalized reports printed in the New York papers regarding the tests at Sing Sing, the commission, when asked if reporters would be allowed to witness their work at Auburn, answered, "Yes, if they would allow themselves to be placed in the chair."[19] Again, a drive belt proved to be faulty, and the commission had to postpone their tests a few hours while it was repaired. As they waited, Warden Charles Durston conducted a tour of the famous prison. They visited the shops, the prison kitchen, and arriving at the mess hall, observed 1,200 "stripeds" at their midday meal. They were escorted then into the chapel where a young black convict entertained them with his vocal imitations of a cornet and bugle.

When the apparatus was ready, the horse and four-week-old calf purchased by the commission were brought in. Similar to the experiments at Columbia and Edison's

---

19   *New York Times*, 1 Jan. 1890.

lab, the work this day consisted of killing animals with electricity. However, in the case of the calf, Dr. Fell immediately set to work with his resuscitation device to see whether the calf was dead or had merely been stunned by the shock. Performing a tracheotomy, Dr. Fell inserted his artificial respiration tubing down the calf's throat and kept its lungs pumping for a half hour as the body cooled and stiffened. But the heart could not be restarted and the commission was in "high glee"[20] over the success of the killing. The nominal reason for Fell's presence was to prove once and for all that the heavy electric shocks caused death and not merely suspended animation. But, as it was well known that Fell would be there with his pulmotor at Kemmler's death, his presence had a more obscure, but more important, symbolic value. Fell, Southwick, MacDonald—the man who'd conceived of the death chair and worked tirelessly to see it through gestation and birth—would be there when it was christened with Kemmler's blood. And that Fell's interest lay as much in restoring life as in taking it made him an excellent witness. A guardian of life, a secular priest, a worker of scientific miracles, Fell—like the other men on the warden's list—would be there to consecrate the event and lend the stamp of civilization's approval.

A week later, the commission was in the northern part of New York, at Clinton Prison, where a third battery of tests was made.

A two-year-old steer weighing a quarter ton was killed there, to the witnesses' satisfaction, and Superintendent

---

20 *New York Times*, 1 Jan. 1890.

Lathrop authorized the payment to Brown of the first half of the $8,160 agreed upon. The second half would come with the first successful use of the machinery on humans.

A little more than two years had passed from the issuing of the Death Commission's report until the approval of the electric chair by Lathrop. Westinghouse had been thwarted; equipment bearing his name was now installed, tested, and ready for use. Defeated in this arena, Westinghouse made a final effort to keep the AC system from being used to kill criminals. Now judges, lawyers, and legal witnesses were pulled into the conflict in a last attempt to keep AC from being confirmed as the "executioner's current."

The New-York Times.

July 24, 1889

TESTIMONY OF THE WIZARD:
Edison's Belief in Electricity's Fatal Force.
He is positive that the alternating current of 1,000 volts would surely kill a man.

The hearing in the electric execution case yesterday was made more than ordinarily interesting by the testimony of Thomas A. Edison. The great inventor walked into Mr. Cockran's office early in the morning accompanied by Mr. Brown and Mr. Kennelly, his chief electrician, smiling and utterly oblivious of the presumed jubilation of Mr. Cochran and the adherents of the Westinghouse Company, who it has been whispered during this investigation, were quite willing to give large sums of money for the privilege of having Mr. Edison cross-examined for an hour or so by the astute Mr. Cochran.

Mr. Edison seemed disposed to offer a little money on his own account. It was announced that he would pay $100 to Mr. T. Carpenter Smith, the witness who testified to having withstood so many and such powerful shocks of electricity, if that gentleman would come out to the Edison laboratory and take 100 volts, alternating current, having the voltage gradually increased from 10 to 100. Before this session was begun another gentleman offered another $100 to Mr. Smith if he would undergo the experiment. As Mr. Smith had testified that on four distinct occasions he received shocks varying in power from 1,000 to 1,500 volts, he will probably jump at this easy way of making $200.

Chapter Six

CRUEL AND UNUSUAL

William Kemmler spent 15 months in custody, almost all of it in the New York State Prison at Auburn. Two steel cells, 5 ½ by 8 ½ feet, had been built there recently, south of the administration building. Next to the cells was the "dance hall," an exercise area 5 ½ by 4 ½ feet. A door led from the dance hall to the adjacent death chamber.

As though there were two Kemmlers—the actual man and the legal test case—he spent the year and a half oddly disconnected from his own fate, oblivious to the machinations and battles that raged in his name. In *Discipline and Punish*, Michel Foucault presents the idea of the prisoner having "two bodies." Just as a medieval king had a double body—the transitory flesh and the intangible "support of the kingdom"—so the condemned prisoner has his corporal form and his symbolic, or figurative, form. Ceremony, expectation, projected wishes and fears, congealed to create a second "body" which to the society at large is far more real and significant than the meat and bones waiting passively to be destroyed. "In the darkest region of the political field," writes Foucault, "the condemned man represents the symmetrical, inverted figure of the king."[1] And when Kemmler took his place in the electric chair, it was like a shadow-king finally taking his place on the throne. Imbued with quasi-mystical power, Kemmler had a unique status, just as the king—unique, supreme, yet trapped by tradition and superstitious dread—is without peer.

In the fifteen months he spent at Auburn, Kemmler was transformed from an illiterate, drunken murderer into a scientific and spiritual avatar. While in prison he was taught to read by Mrs. Durston, the warden's wife. He spent much of his time writing his name over and over again, as if the more often he scrawled the fourteen letters, the more real he might become. Besides writing materials, he was allowed a Bible, and he pored over the illustrations for hours.[2]

While lawyers, judges, and legal experts brought in by the Westinghouse and Edison factions fought his case, Kemmler sat and listened to his guards read popular novels. While his name was spoken and his fate argued—all the way to the United States Supreme Court—Kemmler spent his time secreted away in a small town in central New York. Auburn lies at the eastern end of the Finger Lakes region, at the midpoint in the 250-mile stretch of Erie Canal from Buffalo to Albany. Then, as now, its most noteworthy feature was the prison, brooding in the middle of town like a garrison fortress.

---

[1] Michel Foucault, *Discipline and Punish*, 28-29.
[2] *Buffalo Express*, 5 Aug. 1890.

Whereas Sing Sing was better known and had a certain iniquitous glamour (the term "sent up the river" derives from so many prisoners being sent up the Hudson from New York to Sing Sing), Auburn was in many ways a more significant element in the state's penal system. Auburn, not Sing Sing or Dannemora, was the place where New York's primary prison innovation took place. In actual construction and layout of the prison, and in the world-famous "Auburn system" of control, the facility was thought to be a model of progress and reform.

Established in 1816 as New York's second prison, Auburn was designed and built in stages. By late 1817, the south wing was ready to receive fifty-three prisoners from the nearby counties to aid in the construction of the rest of the facility. The main building and the south wing were finished the next year and contained sixty-one double cells and twenty-eight so-called apartments that would house up to twenty inmates. Further construction followed and the north wing, finished in 1821, consisted entirely of solitary cells. Influenced by the "Pennsylvania system," of total isolation—solitude was thought to give time for reflection and contrition, and to protect naïve criminals from the hardened—and in response to conditions thought to encourage rebellion and abuses of the pardon system, the new north wing housed its prisoners in cells 7 feet long, 7 high and 3 ½ feet wide. The walls were made of stone and the doors of oak planks bound with iron straps.

On Christmas Day 1821, the first grading of inmates was done, separating them into three classes: the most hardened, who were kept in complete solitude with no labor or other outside activity; those who were kept in solitude but allowed at times to work at some tasks; and those who slept in individual cells but who worked with others by day. The grading system allowed prison officials to treat all the men exactly alike—regardless of behavior or record. No special considerations or privileges were allowed.

But because the system of absolute isolation was responsible for multiple suicides and a great increase in mental illness, a new arrangement was developed. This "Auburn system" was to revolutionize the design and administration of American prisons. Based on "humanitarian" and "philanthropic" ideals, the Auburn system's goal was the utter regimentation of the prisoners. The most important characteristic of the system was the obsession with silence coupled to constant surveillance.

Absolutely no communication was allowed among the men. Their eyes were to be downcast at all times. They did the "Auburn shuffle" marching in lockstep with one hand on the next man's shoulder wherever they went. Prisoners were never allowed to be face to face with each other. Even at meals, they sat at special narrow tables to prevent eye contact. In the shops, no speech was tolerated, and at night the guards moved about the galleries wearing only woolen socks, in order to walk noiselessly and listen in on any attempted communication. No letters or other communication from the outside world were allowed and the only reading materials permitted the men were Bibles.

In order to maintain this absolute order, floggings were frequent, brutal, and often arbitrary. One prisoner discovered with a pencil in his cell received the same six lashes as another man who had tried to throw a fellow prisoner off the roof. Another inmate received four lashes of the whip for "grinning darkey fashion."

The combination of violent intimidation and constant activity was designed to promote a change in the prisoner's disposition. Louis Dwight, one of the country's foremost prison reform advocates, was greatly impressed with the system and toured the United States trumpeting its praises. He founded the Boston Prison Discipline Society, and joined with Baptist and Congregational ministers to promote the new way, not as mere good management but as a method of achieving redemption. "It is not possible," he wrote, "to describe the pleasure which we feel in contemplating this noble institution... We regard it as a model worthy of the world's imitation."[3] So taken was he with the Auburn system that he suggested it be applied to other institutions. The extreme order and unceasing vigilance provided a "principle of very extensive application to families, schools, academies, colleges, factories, mechanics' shops."[4]

The Auburn gospel was spread wide, and by 1833, impressed by the appearance of order and safety, thirteen other states had begun building prisons along the same lines.

For a time, Auburn Prison was the most visited

---

3   Prison Discipline Society, *First Report*, 1826, 36.
4   Prison Discipline Society, *Fourth Report*, 1829, 61-63.

tourist attraction in Western New York, after Niagara Falls. The prison was built so that the entire interior courtyard could be seen at a glance. Visitors were not allowed inside the shops, but the wall at the rear had been built to allow for a narrow passageway. Tourists could move through this corridor and watch the prisoners at work through numerous tiny peepholes in the wall. Elsewhere, there were other places where tourists could observe the inmates without their being aware of it. At first, the entry price was 12 ½ cents per visitor. The fee was soon doubled, not as a means to increase revenue but to dissuade people from coming in such great numbers. At a time when Auburn's population was 5,600, between 500 and 700 visitors each month came to peer at the "stripeds" from secret galleries. Gershom Powers, a judge and later warden at the prison, sold a guidebook to tourists for $.25 a copy.

Powers was a believer in the total subjection of the prisoner's will and identity. A fiercely religious man, he was convinced that the Auburn system could instill in the inmates spiritual ideas and feelings. The constant surveillance and activity, the perpetual threat of beatings, the utterly unchanging routine, were maintained to literally dehumanized the prisoner. The goal was to make him into a uniform being with no identity, "a silent and insulated working machine."[5] Penologist Orlando Lewis visited the prison at the turn of the twentieth century and said that it had "the beauty of a finely functioning

---

5   New York Senate Legislative Documents, 69th session, 1846. vol. 4, no. 20, 6.

machine. It had reduced the human beings within the prison to automata."[6]

With the demolishing of the old work-shops in the 1840s, visitors were allowed to come and be seen by the prisoners, who, however, still were required to maintain utter silence and passivity. The presence of women was a particularly vexing problem, leading to masturbation, which the prison officials believed caused hallucinations and mental illness.

The prisoners were a titillating spectacle—half human, half machine, wearing outlandish striped suits and expressionless faces. Their violence, passion, and hatred were suppressed to the point of zombie-like obedience, yet the impulses remained and on rare occasions one of the men would lose control and fling himself at a female visitor. His punishment was of course severe and quick in coming.

The relationship between the audience and the spectacle becomes confused here. Both prisoners and visitors were on display. The Auburn system was an uneasy mixture of entertainment and moral elevation, of religious training and emotional subjection. And it captured the imaginations of not only American tourists but also prison officials and social scientists from overseas. The reputation of the Auburn system—the financial success of its shops as well as the machinelike subservience of its inmates—spread to Europe and South America. In a particularly American arrangement, the Auburn System combined social control, not-so-subtle psycho-sexual

---

6   Orlando Lewis, *The Development of American Prisons and Prison Customs: 1776-1845*, 78.

theater, and baldfaced opportunism.

While there is a long tradition of incarceration as entertainment (especially in mental institutions where "performances" and tours were quite popular), Auburn's allure was not merely that visitors could come and gawk at the dregs of society. The attraction was enhanced by the idea of novelty and progress. Visitors might tour along the Erie Canal, another Western New York success story, stop at Auburn to take in the show, and then move on to Niagara Falls, the harnessing of which would soon be another triumph of civilization.

While the electric chair was used at Sing Sing only a year after its christening at Auburn, and became associated with that more notorious prison, it's fitting that it was used first at Auburn, the site of so much of what was considered progressive and innovative thinking. Though other prisons had abandoned most of the Auburn system by the 1880s, when William Kemmler arrived there, much of the "modern" scheme—by then archaic and discredited, as innovation almost always becomes—was still in place. The lockstep wasn't abolished until 1900. The striped uniforms were replaced in 1904. And the system of silence was still being enforced in 1910. While one Kemmler—the thirty-year-old male body ravaged by alcoholism and months of passivity—waited in Auburn, the other Kemmler—the public icon, juridical subject—was argued over, dissected, and analyzed.

Kemmler's first trial had lasted only four days. He was easily convicted and sentenced to die by electricity. But within a month, his lawyers argued that death in the electric chair was a violation of federal and state prohibitions against cruel and unusual punishment. Cayuga County's Judge Day granted a stay of execution and appointed Tracy C. Becker as referee to take testimony on the question and issue a report to the court.

Kemmler had only a few hundred dollars to his name at the time of his conviction. Where that money went is unknown, but the fact that he had a lawyer of high regard—Charles S. Hatch—for his first trial, and for his appeals an attorney of national reputation—W. Bourke Cockran—indicates that money from outside sources was flowing freely into the case. The hearings and appeals were Westinghouse's last ditch effort in the war of the currents. If he could have electrocution declared cruel and unusual, then his AC would not acquire official status as "the executioner's current." The appearance of W. Bourke Cockran in the case indicates how badly Westinghouse wanted electrocution prevented. Though there was never an admission that he was in the pay of Westinghouse, no other explanation for his involvement in the case is convincing. During the hearings, the question of his connection to Westinghouse was addressed directly again and again, but every time, Cockran sidestepped the issue.

At the time Cockran became involved in the dispute he was one of the best known and most highly regarded lawyers in the country.

He served in the United States Congress during the 1887-89 term, took a two-year hiatus during which he represented Kemmler, then returned to the House of

Representatives and spent ten more years there.

As part of the Tammany political machine, he was one of the most powerful men in New York City. When Tammany boss John Kelly died in June of 1886, the twenty-four leaders of the assembly districts—the executive body of Tammany Hall—decided that there would be no single boss but a ruling committee. Quickly though, factions formed and four leaders took power: Richard Croker as chief, Hugh Grant, Thomas Gilroy, and W. Bourke Cockran. For the next fourteen years, Croker was the undisputed ruler of New York. During this time, Cockran was known primarily for his brilliance as a public speaker and a shaper of public opinion. Called the "silver-tongued orator," Cockran could draft a platform and stir a crowd like no one else. Born in Ireland, he came to the United States in 1871 and was often used by Tammany to fire up New York's Irish population.

The period of Croker's administration has been called the city's darkest age: no public agencies existed to serve the needs of the 18,000 immigrants who entered New York each month, there were no shelters for orphans or street people, no controls on police violence, pimping or prostitution, and only sporadic concern for public sanitation and health. And despite improvements in the ballot laws instituted in 1890, tampering with the vote continued to be the norm. Bribes, kickbacks, blackmail, and favoritism were an integral part of Tammany rule. An investigative commission exposed a great deal of Tammany corruption; nonetheless, the election of 1890 was a landslide for the Democrats. Apparently the voters didn't mind corruption as long as their livelihoods remained secure.

Though he later broke with the Croker administration, there's little doubt that Cockran was deeply involved in the scheming that kept his faction in power. It's in this context that Cockran must be seen: a brilliant speaker (Winston Churchill called him the greatest American orator), a clever exploiter of public whims and prejudice, a man who moved in the highest circles of government. In short, W. Bourke Cockran was the best lawyer money could buy.

The hearings began on July 9, 1889, at Cockran's New York City law office in the Equitable Building. Dozens of scientific witnesses (among them three future presidents of the American Institute of Electrical Engineering: Franklin Pope, Arthur Kennelly, and Schuyler Wheeler), lawyers, businessmen, and reporters crowded the room. The newspapers printed lengthy daily extracts from the testimony, and editorialized often, usually attacking Cockran and Westinghouse as stumbling blocks in the road of progress.

The first witness was Harold Brown. Cockran began his attack the first day by grilling Brown about lightning and comparing it to man-made electricity. The pressure of lightning has been measured at several million volts. With a peak current of 20,000 amps and a temperature that at times reaches 50,000°F, it was an infinitely more powerful discharge than any that would be mechanically generated at the time.

"Yet," Cockran said to Brown, "you have heard of a

human being having been struck by lightning and not being killed."

"Yes," the witness admitted. "I have."[7]

In the course of the next three weeks, Cockran would introduce a parade of people who'd been struck by lightning and though badly hurt, had survived.

Cockran attempted with some success to confuse and embarrass Brown with questions of an abstruse and theoretical nature. Brown, frustrated and angry, finally admitted that he was an expert only in "commercial electricity" and not in electrical science.

Cockran questioned Brown again the next day, this time concentrating on Brown's animal experiments.

> "Did you find it impossible to kill any of the dogs experimented upon?"
>
> "Yes, there was one dog which the highest pressure failed to kill," Brown said. "But we used the continuous current with him. We did not subject him to the alternating current at all."
>
> "Why did you not? Because you thought he was too tough to kill anyhow and you were afraid of exposing your theory concerning the fatal qualities of alternating current?"
>
> "No, sir, not at all. The reason we did not use the alternating current on Ajax—the dog's name, you know—"
>
> "Named Ajax," Cockran interrupted, smiling, "because he defied lightning?"

---

[7] *New York Times*, 10 July 1889.

Brown went on. "It was because we had already used the continuous current and if we had killed him by the alternating current the advocates of the latter system would have said that he had been so injured by the continuous current as to be practically killed and the alternating had nothing to do with his death."

After being questioned about the unpredictable deadliness of lightning, Brown stated that on occasion a person struck in a thunderstorm might survive because "lightning does not follow the ordinary rules followed by currents which are understood." As the most vocal prophet of electrocution, and as builder of the prototype electric chair, Brown was repeatedly questioned about the predicted results of the first electrocution. He claimed that it would be "inevitably fatal" and that death would occur within fifteen seconds, producing no burning of the flesh.

"Have not persons receive the full strength of such an electric shock without being killed?" Cockran asked.

"Not when the contact was good."[8]

And as he did again and again during the hearings, Cockran questioned how anyone could know the actual effects of the electric chair if it had never been tested.

The following day, the issue of the human body's resistance to current was addressed. Franklin Pope, a Westinghouse engineer, was brought in to testify and he disputed the accuracy of the Wheatstone bridge, a device for measuring resistance. He claimed that it only worked

---

8   *New York Times*, 11 July 1889.

well on metals and that the human body, being far more complex than a piece of wire, and possessing a varying set of internal circumstances (temperature, levels of hydration, muscular tension, hormonal activity, and so forth) could never be assessed properly. Again the issue of lightning was brought up the problem of the flesh burning was raised. Edison's people, aware that charring of the flesh would prejudice the public against further use of the chair, insisted that no burning would take place.

District Attorney George T. Quinby, who'd come from Buffalo for the hearings, accused Cockran in his cross-examination of being in the pay of the Westinghouse Company, and claimed that all his efforts were colored by this fact.

According to the *New York Times*, Franklin Pope admitted that Westinghouse "objected to its system being used in electrical executions because such 'exhibitions' would be calculated to make the commercial public regard the system as dangerous in practical use."

"Has the company engaged counsel to urge this objection?" Quinby asked.

The *New York Times* described Cockran at this point suddenly becoming "intensely interested in the architecture of the ceiling."

"Not to my knowledge," Pope replied.[9]

On July 12, their hearings were moved out to Edison's lab in New Jersey. A large entourage of lawyers and reporters arrived there in the morning and were given a tour. Edison's inventions were demonstrated and the

---

9   *New York Times*, 12 July 1889.

visitors were treated to hearing a phonograph recording of a cornet. Then various individuals' resistances were measured with the Wheatstone bridge. On this occasion a great variety of resistances (from 1,300 to almost 10,000 ohms) was noted. In some cases the same subject was tested twice and his resistance was found to have varied greatly within a few minutes.

The events of the day came to a close with another argument and barrage of accusations: this time because one of the visitors brought along by Cockran turned out to be John Noble, a Westinghouse engineer. He was denounced by Harold Brown as a spy and an interloper. Referee Becker eventually settled the hostilities and got the group back to the business at hand. Further experiments with human resistance were performed.

Back at Cockran's office, the next witness was Daniel Gibbons, a man who'd studied electrical shocks for years. He'd witnessed Brown's experiments at Columbia and contrary to Brown's claims, asserted that the animals had suffered greatly.

"What was the effect upon the dogs which were not killed outright," Cockran asked.

"They suffered the most horrible agony. In fact, it was one of the most frightful scenes I have ever witnessed. The dogs writhed and squirmed and gave vent to their agony in howls and piteous wails until at length, exhausted, they fell upon the floor of the cage worn out with their sufferings."[10]

After two engineers testified, a dog named Dash was

---

10   *New York Times*, 16 July 1889.

brought in as an exhibit. It was bound up in bandages, having been shocked by stepping on a telegraph wire. The dog had been thought dead at first, being thrown four feet in the air by the shock. But its owner, thinking to draw the electricity out of it, had it buried up to the neck. It revived and was there at the hearings as further proof that electrical shock, no matter how powerful, did not always cause death.

On July 16, Dr. Frederick Petersen admitted in his testimony that certain of the dogs they experimented on had lived a long while after being shocked. In the case of one, the autopsy found that the heart was still beating thirty-six minutes after the dog had received the electrical jolt.

Referee Becker then asked,

> "In the electrical appliance for executing criminals, are there not many elements of uncertainty, any one of which might result in making the execution a failure? For instance, the belt on the dynamo might slip, parts of the dynamo might become disarranged, the wires might be insulated, the electricity might not be properly applied, and then there's the element of uncertainty concerning the amount of the subject's resistance. Do you think that there are more or less elements of uncertainty about an execution by electricity than by hanging?"

"Well," Petersen said, after thinking a while, "I think that more care would be required in the arrangement for electrical execution than for hanging a man."

"Is hanging painful?"

"Not when the neck is broken. When the criminal dies from suffocation, it is."

"How is death from electricity caused?" Becker asked.

"It is not known to science."

"How is death caused by hanging?"

"By the breaking of the spinal cord or by suffocation."

"Then we do know how hanging causes death. Death by hanging is mechanical. How about electricity?"

"Well, it's mechanical and chemical."

"And because of the tremendous velocity of the electricity, death by that means, you think, is painless?"

"Yes."[11]

Elbridge Gerry finally arrived in New York on July 18, having been on his yacht until then, and took the stand. As in his questioning of Petersen, Cockran again asked how exactly electricity kills. Gerry stated that it does so by "paralyzing the nervous system."

"Don't you think," Cockran asked, "that the term 'paralysis of the nervous system' is all gibberish?"

"I don't know what you mean by gibberish."

"Gibberish, I will say for your information, is the stringing together of a lot of words full of

[11] *New York Times*, 17 July 1889.

importance to the ignorant, but empty in the ears of the enlightened."

"Then I don't think that the term is gibberish at all."

Asked why he was so convinced that electricity would be the best method of execution when over half the respondents to the Death Commission's query thought otherwise, Gerry called on the name of Thomas A. Edison. "I think he knows more about electricity than any other living man."

And grilled regarding the extent of his electrical knowledge, Gerry replied angrily, "My knowledge just about equals yours, and the less you know about it the better you seem to cross-examine."[12] When the hearings began again, Cockran appeared with his hair cut very short and a bandage around his hand. Though he claimed he'd merely hurt himself swimming, it was generally agreed that he'd been experimenting with electricity and had burned himself.

The next few days of the hearings were a parade of doctors who had either treated accidental electrocution victims or had made studies of electricity on animals. Many stories were related of people hit by lightning, including a man who had been struck directly and had his clothes blasted off his body, yet had lived. The state brought in as a witness Dr. Rockwell, "an expert in nervous diseases and electro-therapeutics." Another of the physicians who had assisted Brown in his experiments, Rockwell swore to the certainty and effectiveness of electrocution, but

---

12 New York Times, 19 July 1889.

had little new to add. The hearings by this point had become a kind of counting game, each side trying to bring in as many witnesses as possible to bolster their position.

Thomas Edison's appearance on July 23 promised to make the hearings more substantive. He was, however, at first more interested in childish challenges than actual scientific discussion. The absurdity of the proceedings was augmented by the volume. Being deaf, Edison required that all the questions be shouted at him. The *New York Times* reported that the hearings that day "might have been heard on the street."[13]

Edison explained at some length why he believed electricity would be effective and painless. As to the question of the body's resistance, Edison stated that on the previous Saturday, he and his chief engineer had measured with the Wheatstone bridge the resistance of 250 of his employees. The average was about 1,000 ohms, he said, with little variation. And responding to a question about the burning of human flesh during accidental electrocution, he insisted that this was entirely due to poor contact. He did admit that Kemmler might be "mummified" if it became necessary to apply the current "in its most wicked and aggravating form," for five or six minutes. Cockran suggested that a better term might be "carbonized."[14]

After a long line of electrocution survivors told

---

13 *New York Times*, 24 July 1889.
14 "The People of the State of New York Ex. Rel. William Kemmler, appellant, against Charles F. Durston, agent and warden of Auburn Prison, Respondent," Court of Appeals, 1847-1911. DCCCXCIII.

their tales, the hearings were adjourned in New York. Resuming a week later in Buffalo's City Hall, the hearings included autopsy reports and the testimony of many doctors who had treated electrical-burn victims. The last day of testimony was August 1, at which time Warden Charles Durston of Auburn prison was put on the stand and questioned about Harold Brown's prototype electric chair. The original contract required that Brown be the executioner. But the time set in the contract had passed, and Durston now believed that he would be the one to pull the fatal switch.

After two sessions in Buffalo, the hearings ended. Stenographers had taken down 350,000 words of testimony during the four weeks of nearly daily sessions. The report finally submitted to the court was 1,025 pages long.

Two months later, Kemmler's lawyers appealed his conviction on the grounds that electrocution was cruel and unusual punishment. From the legal firepower brought in by the state to fight the appeal (the attorney general, the deputy attorney general, as well as two district attorneys) it was clear that New York had determined that its new law would not be declared unconstitutional.

In his opinion, Judge Day of Cayuga County Court first addresses the issue of whether the law is within his jurisdiction. After he affirms that it is, he summarizes the arguments for and against. Cockran argued before the court that by undergoing electrocution Kemmler would suffer "the most extreme and protracted vigor and subtilty [sic] of cruelty and torture." Also, the fact that no one had as yet been executed in the electric chair made Kemmler an experimental victim. The state's lawyers, on the other hand, contended that the new method was "one promotive of reform, a step forward and in keeping with the scientific progress at the age." Interestingly, both Cockran and the state's lawyers based their arguments on the grounds of mercy and humanity.

And though over 1,000 pages of testimony were delivered to the court, Judge Day does not discuss any of the facts presented. For him this was entirely a procedural question: whether the court had the right to declare electrocution unconstitutional, basing its decision on law rather than fact. Every presumption, Judge Day declares, must be in favor of the state. The burden was on Kemmler's lawyers to prove that the electric chair was cruel and unusual.

Judge Day asks whether "in this case it has been plainly and beyond a doubt established that electricity as a death-dealing agent is likely to prove less quick and sure an operation than the rope." He declares that "the question is... largely one of fact." Yet knowing that the case would be appealed again and addressed by higher courts, he refuses to deal with the factual evidence presented to him and reaffirms the judgment rendered by the original judge and jury.[15]

Cockran immediately appealed to the Supreme Court of New York. Writing for the three-man panel, Judge Charles Dwight discusses first the history of the prohibition against cruel and unusual punishment,

---

15 *New York Supplement*, vol. 7, 146-52.

describing which methods of execution manifestly fall within this category. Judge Dwight traces the roots of the United States Constitution's Eighth Amendment back to England in 1688, when the "Bill of Rights" (actually called "An Act of Declaring the Rights and Liberties of the Subject") was formulated. He notes that the prohibition against cruel and unusual punishment was not directed so much against the government in its legislative capacity as against the gross abuses of discretion by individual judges. However, regarding the New York law, he states, "It would seem that the provision in the state constitution against cruel and unusual punishment, if it were to have any practical operation—if it was anything more than a mere glittering generality calculated to please popular fancy and gratify the popular taste for declarations of rights—must have been intended as a restriction upon the legislative authority."

Of the methods that clearly are cruel and unusual, he mentions burning at the stake, breaking on the wheel, being fired out of a cannon, hanging in chains to die of starvation, disemboweling and crucifixion, then compares these in a rather cursory manner to electrocution. He concedes that the electric chair is indeed unusual, in that it had never been used before, but then claims that "there is no common knowledge or consent that it is cruel; on the contrary, there is a belief more or less common that death by electric current, under favorable circumstances, is instantaneous and without pain."

He then moves on to discuss the scientific background to the case, though never addressing any of the facts presented in July. The massive volumes of testimony were like an embarrassing relation, shunted to the back room, mentioned but never treated directly. All the appellate judges, though paying lip service to scientific fact, appear to have had a strong desire to take the question on a purely procedural or philosophical level.

Again affirming the constitutionality of electrocution, Judge Dwight brings his opinion to a close with the following discussion:

> It detracts nothing from the force of the evidence in favor of this conclusion that we do not know the nature of electricity or how it is transmitted in currents, nor how it operates to destroy the life of animals and men exposed to its force. Neither do we know the nature of the attraction of gravitation, which is the operative force employed in the infliction of death by hanging, nor how that force operates to draw toward each other masses of matter freely moving in space. We know these, and all other forces of nature, by their effects, and we avail ourselves of them in the daily processes and pursuits of life with a confidence based upon common experience, without inquiry to the essence of the force, or the mode of its operation.[16]

Though there is no evidence that the two men ever met, Kemmler was represented a third time by W. Burke Cockran. He took the case to the highest court in New York State, the Court of Appeals. After reviewing the

---

16   *New York Supplement*, vol. 7, 813-18.

case—historical background, jurisdictional and procedural questions—Justice O'Brien concedes that "the infliction of a death penalty in any manner must necessarily be accompanied with, what might be considered in this age, some degree of cruelty." But he states that the case didn't rest on the factual nature of the question—whether electricity caused extreme pain and lingering death. This, he believed, had already been settled by the extensive research performed by the legislature's Death Commission. "The determination of the legislature of this question is conclusive upon this court." In other words, the question had already been raised and answered. Justice O'Brien, like his predecessors on the lower courts, disregards the testimony from the hearings, calling the documents "a valuable collection of facts and opinions... but nothing more."[17]

Simultaneously, an appeal was made by another lawyer, C.W. Sickmon, on Kemmler's behalf, arguing that he'd been mentally irresponsible for the murder because of extreme alcoholism. A Judge Gray gives some attention in his opinion to the statements made by doctors who had examined Kemmler, and to Kemmler's mental state after the murder, and finds that on all the questions raised by Sickmon, the courts had in fact acted properly and that Kemmler's conviction should not be overturned.

Curiously, he notes that though the question of cruel and unusual was being at the same time addressed by another member of the court, and was therefore not within

---

17   *North East Reporter*, vol. 24, 6-9.

his purview, he nonetheless felt compelled to make his opinion known. "Punishment by death, in a general sense, is cruel; but as it is authorized and justified by a law adopted by the people as a means to the end of better security of society, it is not cruel, within the sense and meaning of the constitution. The infliction of the death penalty through a new agency is, of course, unusual." But even though this is the case, Justice Gray goes on to state that the enactment of the legislation was based on factual investigation and that Kemmler's appeal was based on only "possibilities and guesswork."[18]

After the third appeal failed, it seemed Kemmler's death would occur shortly. In the prison shop at Auburn, two convicts put the finishing touches on the pine coffin that would hold his remains. And Kemmler made out his will, a process that the popular press delighted in describing as childish and farcical. He left the few things he owned to those he'd spent time with in prison. To Daniel McNaughton, one of his keepers, went his pictorial Bible. His testament was given to Bill Wemple, another guard at Auburn. He gave his pigs-in-clover puzzle (a one-person game played with marbles to be tilted into holes on a board) to Referend Dr. Houghton, and his slate, covered with his autographs, to Mr. Yates, the prison chaplain. The primer from which he had learned to read was returned to Mrs. Durston, the warden's wife, as was his little book of Bible stories. In addition, he gave Mrs. Durston fifty cards with his autograph laboriously written on each.

---

18   *North East Reporter*, vol. 24, 9-11.

A statement attributed to him was released by the prison authorities:

> I am ready to die by electricity. I am guilty and I must be punished. I am ready to die. I am glad I am not going to be hung. I think it is much better to die by electricity than it is to be hung. It will not give any pain. I am glad Mr. Durston is going to turn the switch. He is firm and strong. If a weak man did it I might be afraid. My faith is too firm for me to weaken. They say I am not converted. I don't care what they think. I know what I've got. I am happy to die. I have never been so happy in my life as I have been here.

Electrician Edwin Davis busied himself in the room next to Kemmler's cage, getting the death apparatus in final readiness. It had been redesigned by Davis after Brown's chair was deemed unworkable months before.

Witnesses had gathered in Auburn, expecting that the execution would occur within a day or two of the final appeal. Among them were Dr. George Fell, ready to test his resuscitation device on Kemmler, and Dr. Southwick, who made this statement regarding the cumbrous straps, wires, belts, and fasteners that were needed to keep the prisoner in the death chair:

> I am opposed to so much paraphernalia, but the present arrangement will have to do, because we cannot afford to suffer failure. The whole world is watching the result of this experiment, and if we neglect any

precautions there might be a slip, and the system would therefore be condemned. I am fully convinced that Kemmler's death will be instantaneous. There is no danger that the current will be kept on long enough to burn his body. I anticipate no disfigurement at all. Every effort will be made to make a positive contact. I would like to have him talking when the current is turned on, for then the attendants could have an opportunity to see how quickly death would follow an electrical shock as Kemmler will receive.[19]

The witnesses had been told to report that night at 9:30, in order to sleep within the prison walls, a tacit statement that Kemmler would be killed the next day. However, a law clerk appeared in Auburn with a writ of habeas corpus issued by Judge Wallace of the United States Circuit Court, ordering the warden to produce Kemmler in court three weeks hence. A new lawyer had entered the case on Kemmler's behalf—Roger Sherman, a former United States assistant attorney. When questioned, Sherman denied that Westinghouse had engaged his services, but refused to say who had paid him to secure the writ that gave Kemmler two more months to live.

Sherman, too, appeared in Auburn. Denied permission to see Kemmler, he presented legal papers and rushed out of town on the 3:05 train. Two hours later, when the warden had broken the news to the angry and

---

19   *New York Times*, 29 Apr. 1890.

frustrated men who had come from all over New York to see the execution, a calf was brought into the death chamber and killed with the machinery that had once again been denied its destined victim. The sacrificial calf was killed with 1,200 volts, and George Fell tried unsuccessfully to bring it back to life.

On May 21, 1890, the case reached the United States Supreme Court. W. Burke Cockran was no longer involved. Roger Sherman attempted to show that electrocution was a violation of United States Bill of Rights, being cruel and unusual punishment and denying Kemmler his life without due process of law.

Chief Justice Melville Fuller wrote the opinion for the court, turning down Kemmler's appeal. In it, he states that "the punishment of death is not cruel with the meaning of that word used in the constitution. It implies there something inhuman and barbarous—something more than mere extinguishment of life." Regarding the procedural issue, Justice Fuller wrote, "In order to reverse the judgment of the highest court of the state of New York, we should be compelled to hold that it had committed an error so gross as to amount in law to a denial by the state of due process of law to one accused of crime." He states "with no hesitation" that this is not the case.[20] The writ of error was denied and the last obstruction to Kemmler's execution had been removed.

A *New York Times* editorial five days later expresses some of the anger, frustration and indignation the authorities felt at the long and labyrinthine appeal process.

---

20   *Supreme Court Reporter*, vol. 10, 930-34.

Calling Cockran and his associates "hurtful," "silly," "frivolous," "putative council," the editorialist rails on at great length regarding the outrages perpetrated on the judicial system by the Westinghouse lawyers. The "repeated raising of frivolous 'points' which the courts consent seriously to entertain must bring the courts themselves into popular contempt." The writer expresses his anger that the Kemmler case should drag on so long, that a business conflict should bring the courts shame and create the public perception that they are merely the battlefield on which the powerful might fight for profit and power:

> Thus far the case has been deeply discreditable to the State of New York. Its judicial machinery has been brought into contempt by the showing that the courts were powerless to protect themselves against the efforts of persons who had no real standing in court to delay justice. If Kemmler were a millionaire he could not have commanded more persistent or unscrupulous efforts to get him free of punishment than have been made at the instigation of persons who do not care a straw whether he lives or dies, and would not give a dollar to save his life from being taken by another agency than that which has been chosen for the purpose under the law.[21]

Though Westinghouse's schemes might be seen as highly cynical, without actual concern for deeper legal or

---

21   *New York Times*, 24 May 1890.

humanitarian issues, Edison and his forces were no more ruled by true concern for morals or ethics.

Courts had argued over the issue for centuries, yet there was a broadly accepted notion of what constituted cruel and unusual punishments: those that go beyond or vary from the current idealized concepts of human dignity, the advancement of culture, decency, and civilization. The United States Eighth Amendment was written as an instrument to separate a barbarous "them" from a civilized "us." Though cruel and unusual is a concept that has traditionally not lent itself to precise definition, the needs of the state, the aspirations of the society, the desire for order and safety, have usually prevailed over more humanistic impulses. Courts frequently depend on the idea of community standards as a test for the Eighth Amendment. Only those punishments that shock the moral sense of the community, or are anathema to "all reasonable men" are forbidden as unconstitutional. Of course what defines community, "reasonable" men, or even the idea of moral sense is difficult to determine. It's not so much the severity or duration of the punishment that makes it unacceptable, but the degree to which it jibes with the notion of cultural advancement. Flaying, crucifixion, dismemberment, and burning are cruel and unusual in the United States not because they are painful or injurious to human dignity, but because they are not informed by underlying values and traditions of American culture. Americans wanted very much to separate themselves from their cousins in Europe. The Eighth Amendment was and is a tool to position the

United States on the leading edge of progress. What is alien or obsolete is much more readily condemned than what merely causes great pain or indignity.

By May of 1890 all obstacles had been removed from the path of electrical death progress. All legal maneuvers had failed: electrocution was accepted as neither cruel nor unusual.

Kemmler the legal case was dead, yet immortalized in court documents—a paradigm, a precedent. Kemmler the man had two months left to live.

The New-York Times.

AUGUST 2,, 1889

NO LONGER INDIFFERENT
KEMMLER AT LAST APPRECIATES THE GRAVITY OF HIS SITUATION

Recent developments have more than confirmed the truth of the statement that Kemmler, the condemned murderer, is weakening daily. The abandonment of all hope and the consequent realization of his utterly lost condition never came to him with such terrible force as they did last night. All night long the vision of his impending fate arose before his sleepless eyes, and as he tossed and tumbled on his narrow cot he cried aloud in his despair, "I wish it were over." His lack of power to control his fears was pitiable to behold.

The cause of this sudden and unconquerable despair was the noise of preparation in the adjacent room of execution, and which the prisoner could not help hearing. Men were making a test of the apparatus which is to send the fatal current through the body of Kemmler. In the room were Warden Durston and Electrician Barnes of Rochester and a few others who had been invited to witness the test.

Chapter Seven
THE EXECUTION

Kemmler sat on the edge of his bunk. Unlike the 1,200 other prisoners at Auburn, Kemmler was allowed by the warden to wear a suit and tie in his cell. He sat; he listened to the sounds from the adjoining room. The clank of wrenches dropped on the stone floor, soft curses as a screwdriver slipped from its notch, a sudden spill of nuts and bolts. A long pause, then a harsh humming and a sudden buzzing sputter. The date was August 1, 1890.

"They're getting ready, ain't they?" he said.

Daniel McNaughton, one of the two men who watched over Kemmler night and day, nodded.

The voices from the next room were unfamiliar. Kemmler heard strange words—"volts," "amperes," "ohms," "armature," "voltmeter" and "insulation"—and understood none of them. The unseen men might have been foreigners, speaking an alien tongue.

That day C. F. Barnes, an engineer from Rochester, had arrived at the prison to begin the final tests and modifications on the death chair. Barnes, who was rumored to be the man who would throw the fatal switch, was sneaked into the prison to avoid the stares and questions of the reporters who thronged Auburn, now that the last appeal had been rejected. He performed adjustments, installed new drive belts on the dynamo's steam engine, and got the bank of twenty-four incandescent lights glowing. All that remained was the replacement of a faulty voltmeter. Edwin Davis had gone to New York City to get a new one.

Warden Charles Durston came into the room where Kemmler's cage stood. He made small talk with his prize prisoner, and told him that his wife, who had been visiting Kemmler every day, was out of town. He wouldn't be seeing Mrs. Durston again. Everyone, even Kemmler, knew this meant the execution was not far off. Four months before, when Kemmler's death had only been prevented by last-minute legal wrangling, Mrs. Durston had also left Auburn to visit her family.

After the warden was gone, the gas lights in the room—which would be replaced later that year by electric lamps—were shut off and Kemmler lay for hours awake, listening.

The newspapers the next day, August 2, began reporting that Kemmler had become unhinged by the long wait. The articles called him demented and brought up the possibility that the warden would need to talk with the governor about going ahead with the execution.

There was little doubt that Warden Durston would have been pleased to be relieved of the responsibility. Since the beginning of Kemmler's stay in Auburn, Durston had stated that he hated the idea of being the first man to supervise the electrical killing. After the execution—with the accusations and recriminations flying—Durston's record was publicly scrutinized for explanations of the "botched" event. Called "Governor Hill's mouthpiece in the vicinity of Auburn,"[1] Charles Durston was more a politician than a criminal justice bureaucrat. For two decades before the execution, Durston was thought of by his enemies as a "political disorganizer and trickster." Having fled the Republican party, he soon became a prominent force in the Democratic machine of Cayuga County.

A clever and determined political infighter, Durston took advantage of any opportunity he could for the advancement of his career. At the Democratic convention in Saratoga, when the party was choosing its candidate for governor, Durston bucked the men higher up in the organization and refused to support Roswell P. Flower, the designated candidate. Durston backed David Hill instead. Hill won the nomination and was sent to the governor's mansion in 1887. Not long after, Durston was repaid for his support by being named warden of the prison at Auburn. Cayuga County Democrats were initially displeased with the choice, and then infuriated when Durston instituted a policy of hiring only his backers as employees in the prison. With David Hill's aspirations for the United States presidency well known, it's not at all unlikely that Durston assumed that he, too, would rise if and when Hill ascended to national prominence.

Durston's motto was "My prison, my convicts." Many

---

1   *New York Times*, 18 Aug. 1890.

charges of mismanagement were leveled at the Durston regime, and his lack of interest in the "moral qualities of his henchmen" was well known. On three occasions, the governor promised Cayuga County Democrats that he would remove Durston, but he neglected each time to keep his vow. When pressed by Auburn Democrats who were clamoring for his removal, Hill finally said he had no authority to dictate the actions of state prison superintendent Lathrop.

It is against this background that the warden's actions during the last week of Kemmler's life must be seen. A man with political aspirations far beyond supervising a prison, Durston certainly would not want the stigma of being the world's first electrical executioner. Many speculated that he hoped Kemmler would commit suicide, putting an end to the long imbroglio. But in fact a strict watch was placed on Kemmler to prevent this. Chaplin Yates of the prison, along with Reverend O. A. Houghton visited frequently, hoping to keep his spirits up. During the previous months, Kemmler had made a large ball of tinfoil from the tobacco given to him. He worked and polished the ball until it was the texture of glass. There was some concern that he might use the ball to kill himself—swallowing it to cause suffocation. However, on her last visit to him, Kemmler gave the ball as a gift to Mrs. Durston.

On August 2 or 3, Durston stopped at the New York offices of Doctor Edward C. Spitzka[2] to discuss the

---

2 The name Spitzka became closely associated with forensic medicine and highly publicized trials. Edward Charles Spitzka, the father, was a notable witness in the trial of Charles Guiteau, assassin of President Garfield. Edward A. Spitzka,

autopsy mandated by the electrical death law. Besides visiting the famous forensic specialist, it's also likely that Durston conferred with Harold Brown in New York regarding the final adjustments to be made on the death machinery. By the time Durston returned to Auburn, rumors, exaggerated reports and blatantly false stories had begun to obscure the situation. The town was overrun by newsmen who, though there was little yet to report, still felt compelled to fill pages. Any piece of evidence was quickly made into an "important" story. A policeman saw two hacks pull up to the prison gate near midnight and told the *New York Times* that this indicated the execution would occur that night, August 3 between two and four AM. The hacks had come from Port Byron, a town on the main rail line between Rochester and Syracuse, and the men arriving were probably there on the business related to the execution. But the policeman's prediction, like many others, proved false.

The actual cause of the delay was Warden Durston's unwillingness to be saddled with the unenviable task of setting the exact time and date of Kemmler's death. He sequestered himself inside the prison on Sunday, when no visitors were allowed. But reporters had stationed themselves around town, along likely routes he might take on the way home, and one newsman was able to wheedle a brief interview with him. Questioned about his trip to New York City, Durston claimed he had

---

the son, performed the autopsy and brain examination on Leon Czolgosz (also killed in Auburn), assassin of President McKinley.

merely gone to discuss matters pertaining to the manufacturing business of the prison. He denied having had any contact with Governor Hill of late or that Kemmler had become an "idiot" while waiting for the end.

The next day, the warden finally sent out the invitations to the execution and presented the list of guests to the press, gathered at Osborne House, a large four-story hotel. Saying as little as possible, his surliness barely concealed, Durston took the reporters to task for all the false and exaggerated stories they'd written. Refusing to answer any questions, Durston stormed out of the hotel.

At this point, it became known that C. F. Barnes had been told via telegram not to bother coming to Auburn that night. Barnes had allowed himself to be interviewed by the *Rochester Post Express* and this had apparently wounded the vanity of the warden. Barnes described the electrocution of a calf at Auburn this way:

> When we were fixing the electrodes on him, he struggled and kicked and there was an air of wonder on his face. When we turned on the current he died so quickly that not a change took place in his features. His eyes did not close nor a muscle quiver and I could scarcely believe he was dead. When we took the current off, he wilted as if every bone in his body was broken.[3]

Again we have the process laid out in terms of sacrifice and crypto-religious awe.

Kemmler's metamorphosis from "hatchet fiend" to

---

3   *Buffalo Express*, 5 Aug. 1890.

willing victim was, by this point, all but complete. As a reward for his beast-like tractability he was allowed on August 4 to visit with Frank Fish, another convicted murderer waiting on Auburn's death row. The two men had occupied adjacent cells and, with Kemmler's death looming near, were permitted to spend several hours together in the large room which their cages opened into. Fish brought his banjo and while Kemmler perched himself on a table in the middle of the room, Fish played airs on his instrument and even sang a few songs. Kemmler nodded, tapped his foot to the music and whistled along, seeming to enjoy himself in a distracted way. The guards stood by listening too, as Fish plucked out "My Old Kentucky Home" and "Wait Till the Clouds Roll By." Kemmler's apathy returned when he was placed back in his cell. This state was not changed by the receipt of a registered letter from his brother, in Philadelphia. He broke the seal and laboriously worked his way through it, showing no sign of interest or real understanding. Then he folded the letter, put it in his pocket and neither mentioned nor discussed its contents with his keepers.[4]

His spiritual mentors, Reverends Yates and Houghton visited him for more than an hour. His prayers and Bible-reading were enough to spark the rumor that flashed through the press corps: that Kemmler's calm had collapsed and he was tormented by visions of the hell that awaited him. According to the rumor, he'd been shot full of morphine to bring him down from his hysterical state. Newspapers disagreed on almost all the facts, printing utterly contradictory

---

4 *New York Times*, 5 Aug. 1890.

stories. Kemmler ate well, he ate nothing at all; he was calm, he was frantic with visions of eternal agony; he slept hardly at all, he slept like a baby; his conversion was authentic, it was just one more ploy in hopes of "cheating the chair." Fourteen months before, the press had made him out to be a brutal, miserable wretch. With his death days or perhaps hours away, they used him as a blank slate on which anything, true or false, could be written. Like the chair, like electricity itself, he had become something the popular mind could project its fantasies onto.

As the time came closer, the press as well as the general population felt the momentum building. The Osborne House swarmed with reporters, and a special Western Union suboffice was set up in the freight depot across the street from the prison, with fourteen additional wires leading to New York. Increasingly, crowds of townspeople pressed their faces against the heavy iron bars of the prison gates, curious, hoping to get a glimpse of the mystery inside. Men, women, and children stood quietly at the gates all day, and the insufferable August heat added to the air of uneasiness.

On Tuesday, August 5, it was clear to everyone that short of intervention by the governor, Kemmler would die the next day. The witnesses had gathered, Kemmler had received the sacrament of communion according to the Methodist rite, and those who were charged with "the duty of ministering to his spiritual welfare" declared they'd done all they could "with one of Kemmler's low order of intelligence."[5]

---

5   *New York Times*, 6 Aug. 1890.

Edwin Davis had appeared in Auburn that morning, carrying with him the replacement for the chair's defective voltmeter. Immediately upon his arrival at the prison, he began work installing the new device.

Questioned by Kemmler about the noises in the next room, McNaughton told his charge that the men were moving equipment from one room to another. For reasons still not clear, Durston had decided on the day before the execution to shift the location of the death chair from the place where it had performed tests perfectly to a spot in the keepers' mess room. Soon after Kemmler's death, a rumor circulated claiming that Durston had moved the chair in order to make the execution a purposeful botch. It was argued that the warden was paid—perhaps by the Westinghouse people—to switch the entire electrocution setup from one place to another in order to facilitate the most gruesome execution possible.

Kemmler spent his last day much as he had spent the previous four hundred: pacing the cell, lying on his bed, making small talk with his keepers. The papers said that he told McNaughton he hoped "to die like a man," bringing no shame on those who were instrumental in his death. The warden came to him on the morning of August 5 and it was reported that he told Durston, "I am not afraid, Warden, so long as you are in charge of the job. I won't break down if you don't."[6]

Before Frank Fish was taken elsewhere in the prison, he was allowed to see Kemmler one last time. The two men shook hands and Fish told Kemmler to keep his

---

6   *New York Times*, 6 Aug. 1890.

courage up. "It will all be over soon. I follow you after a little while," he said, as though they were explorers heading into unknown territory.

At three PM all the witnesses[7] went en masse to the prison and were shown the new death chamber by the warden. The chair, in its final form, was square, stout, and straight. Made of heavy oak timber, there was not a curved line in it.

An extended leg rest—which made it resemble a chaise longue—was part of the original design. But it was removed. And though some contemporary illustrations show Kemmler's legs in a horizontal position, he was in fact killed with his feet flat on the floor.[8] The upright crossbars for the back, the arms, rungs, and legs were all crude lumber, and the entire device sat like a ponderous throne bolted to the floor and insulated at the feet. The arms were wide. The seat was perforated wood. The upright posts of the backrest rose higher than the condemned man's head and from the upper crosspiece against which his head rested, a wooden structure like a miniature gibbit extended. A quarter-inch hole was drilled through the horizontal portion of this "gallows," through which the heavy wire stem

---

7   Dr. Lewis Balch, Dr. W.T. Nelson, Dr. J.M. Jenkins, Dr. Joseph Fowler, Dr. Henry Argue, Dr. C.W. Daniels, Dr. A.P. Southwick, Dr. H.E. Allison, Dr. E.C. Spitzka, Dr. Carlos MacDonald, Dr. George Fell, Oliver Jenkins, Joseph Veiling, Horatio Yates, Tracy Becker, Michael Conway, George Bain, Frank Mack, Dr. George Shrady, George Irish, Dr. W. T. Jenkins.

8   Carlos MacDonald, *Report of Carlos MacDonald, M.D., on the Execution by Electricity of William Kemmler, alias John Hart*, 6.

of the head electrode passed. A similar hole was drilled in the lowest crosspiece for the spinal electrode. The overhanging brace was movable up and down through iron collar clamps on the back of the headrest.

Broad leather straps bound the prisoner's wrists and elbows to the armrests. Buckled belts on the side posts of the backrest pinioned his upper arms and circled his waist. Attached to the headrest was a pad, heavily insulated with rubber, designed to fit the curves of the neck and upper spinal column. The prisoner's head was drawn back, hard and tight, into this neck saddle by a leather mask that covered the forehead, eyes, and chin, but left the mouth and nose exposed.

The electrodes were flexible rubber suction cups about four inches in diameter. They held a perforated copper disc that was attached to the wires. A sponge inside the cup, capable of holding a quarter pint of potash solution, would be the ultimate conductor of the current to the prisoner's body. A tension spring pressed the sponge against the skin.

As the witnesses clustered around, Dr. George Fell seated himself in the chair and allowed himself to be strapped in. After a low level current was run through his body, he pronounced the chair ready for use. The image of Dr. Fell strapped in and tasting a mild version of what Kemmler would soon experience might strike one as odd or grotesque, like a little boy playing with a new toy. But all of the men who had worked so hard to make the chair come into being on some level thought of electricity as a kind of numen—creative force and unknowable danger.

Returning to the Osborne House that evening, the witnesses said they were still unsure when the execution would take place. And much to everyone's surprise, C.F. Barnes had arrived, though Durston had told him to stay away. Barnes asked if he could speak privately with the warden. They went off together and five minutes later came back all smiles, their differences apparently settled.

There was a flurry of excitement that evening. About seven PM the witnesses started from the hotel and walked as a group down the street to the prison. The people of Auburn were aroused quickly and soon there was a large crowd following the procession, sure that the final hour had come. The prison's iron gates swung open, the phalanx of bearded and dark-suited men entered and the gate swung shut. The crowd grew in size until there were nearly a thousand men, women, and children gathered outside the prison walls in a death vigil. The crowd was surprisingly quiet. Those who spoke, did so and whispers, "as though a feeling of awe had settled upon them."[9]

The crowd waited a half hour, seeing only an occasional glimpse of the doctors behind the grated windows. Word quickly spread that the execution was over and the crowd began to disperse. Those few who lingered outside the prison found out later that the witnesses had merely gathered to discuss the details of the autopsy and general post-execution arrangements.

The witnesses, returning to the Osborne House, left word that they should be awakened at 4:30 AM.

Spectacle and secret, revelation and ritual: the

---

9   *New York Times*, 6 Aug. 1890.

execution was full of contradictory impulses. The day before the killing, Dr. Southwick voiced his utter opposition to the idea of secrecy in the execution and autopsy. He told newsmen that every detail should be made public because the success or failure of the system depended on public knowledge and acceptance. The average citizen, Southwick said, had the right to know the exact results of the killing, and even though the law prohibited any reportage of the actual event, as far as he was concerned, the public would know. Again he expressed his hope that electrical execution would become universal and stated that he intended to go to Paris the next year to introduce it there.

Hidden spectacle: a paradox that illustrates the needs driving the execution. The men behind the electric chair required public confirmation that it did all they claimed for it. And yet they also felt the need to keep the execution hidden so as not to excite the base, crude, vulgar displays that often accompanied public executions. The electric chair was to be an instrument to edify the young regarding the consequences of crime and to demonstrate the "majesty of the law." Kemmler's death would be a public burning, partaking of both mass affirmation and inquisition-style secrecy. Again the idea of the two Kemmlers comes into play. A public and a secret Kemmler were killed: the childish, dim-witted little man and the exemplar, the scapegoat, who died in order that science and society move forward.

At four o'clock the next morning, townspeople were already filling the streets of Auburn. The day before had been unbearably hot, and the temperature overnight had dropped only a few degrees.

Among the streams of people spilling into the pre-dawn streets were men who'd slipped out of the Osborne House in twos and threes. Reaching the prison's high iron gate, they showed their passes.

The guard on duty nodded and opened the gate to let them enter.

Shortly, still well before the sun was up, the streets were crowded with those who'd come to be near when Kemmler was put to death. On telegraph poles and trees, young men perched to get a better view over the forty-foot stone wall. On roofs and in second-story windows of nearby houses, the faithful had gathered for a glimpse. By six AM, there was a crowd estimated between 500 and 1,000. Those who'd arrived early were pressed up against the iron bars by the crowd behind them.

A group of guards, coming off the night shift, emerged from the prison and marched in single file through the gate. The throng pulled back to let them pass, and the men were pestered with questions as they went by. None of them, however, knew a thing about Kemmler. None of them had seen Kemmler since he'd arrived months before.

Suddenly, a bell was rung inside the prison and the lookouts waved their handkerchiefs as the signal that it was over. A cry of "He's killed, he's killed!" went up from the crowd. But again rumors, expectations, and hopes were proved wrong. The bell had merely been a signal for the civilian employees to fall in line to enter the prison. The iron gate swung open and the crowd drew back to let them by.

Inside the walls of the prison, hidden from the thousands of eyes, from the voices and the already sweltering heat, Kemmler sat on the side of his bunk. Hearing footsteps, he looked up. By the light of a hissing gas jet he saw Reverends Yates and Houghton, the warden and J.C. Veiling, a deputy sheriff who had been friendly with Kemmler while he was in jail in Buffalo.

Durston entered and read Kemmler's death warrant. The condemned man's only response was, "All right, I am ready." Kemmler, displaying the same sang-froid he had maintained for weeks, insisted Veiling stay for breakfast. Veiling agreed, and while they waited for the food, the two clergymen entered the cell. They talked briefly with Kemmler, then knelt and prayed with him. Breakfast arrived and the condemned ate hungrily, though his dining partner found it difficult to get much of the food down.

After Kemmler had cleaned his plate, he was told that he would need to have his hair cut. He had taken great pride in his hair, which was dark brown, wavy, and fell in a Hyperion curl across his forehead. He'd combed and recombed it that morning, and now asked if the haircut was necessary. The warden insisted. Veiling had come equipped with clippers. He was nervous, and did an amateurish job, cropping a spot 2 ½ by 1 ½ inches on top of Kemmler's head. The cut looked like a great scar, but Veiling did manage to save the curl.

While he was being worked on, Kemmler spoke with Veiling. "They say I'm afraid to die, but they will find that I ain't. I want you to stay right by me, Joe, and see me through this thing and I promise I won't make any trouble."[10]

Seeing that things were going well here, Durston left the cell and in the entrance hall above met the men who had accepted—and in some cases pulled strings to receive—his invitations. Yet the hour appointed for the witnesses to arrive had come and gone and still some were missing. The warden's frustration was obvious as the minutes ticked away. It had been decided that the execution must take place after sunrise as daylight was essential for the autopsy. However, the dynamo that powered the chair was used primarily in the prison shops between seven AM and five PM If the dynamo were stopped during working hours all 1,200 prisoners would know that the execution was occurring. And the warden didn't want to find out what their reaction would be to that knowledge.

"Gentlemen, I will not wait any longer for those who are not here," the warden said, looking a last time down at the gate where the crowd continued to swell. "This affair cannot be made subject to personal convenience, and I think it unfair to me that I should have to be kept until this hour."[11]

Silently, the witnesses followed Durston to the rear hall where a guard stood ready to open the entrance to the basement. A huge iron-bolted door swung back and the warden led the way. The men descended an iron stairway to the stone-floored corridor, then stooped to pass

---

10   *New York Times*, 7 Aug. 1890.
11   *Buffalo Express*, 7 Aug. 1890.

through a doorway in the massive basement wall. Soon they came to a room faintly illuminated by early morning sun. The death chair loomed in the shadows, becoming steadily more solid, more distinct, as the wan light streamed through the windows.

The room was empty, cold, and barren. The voltmeter, switches, light-board, and other instruments for controlling the current were in an adjacent room. The dynamo[12] and engine to run it were located a thousand feet away in the loft of the prison marble shop. Heavy wires led through a small hole in the shop wall, up to the top of the building, then traveled along the roofs over the main wing and then crept among the vines and runners that covered the face of the administration building. The wires entered the basement through a square window, covered with bars and an iron grate, and ended the journey in the control room.

Only a few minutes before Kemmler was brought into the room, an important question remained to be answered. The warden asked the two doctors in charge, Drs. Southwick and Spitzka, how long the current should be left on.

The doctors looked at each other, silently, unsure. Finally Dr. Spitzka said, "You have more experience in

---

12 A 50-horsepower engine, belted to a two-inch shaft, drove the dynamo. On the shaft were a 36-inch pulley run on a belt to the 12-inch dynamo pulley. Also, a 30-inch pulley was belted to the exciter. The dynamo's specifications were as follows: commercial voltage, 1,600; mean voltage, 1,512; maximum voltage, 2,376; revolutions per minute of the armature, 1,500; alternations of the current, 230 per second.

these things than I do."

But Dr. MacDonald's expertise was only with dogs and cattle, not humans. "10 seconds at the least," he said. "But I assume more will be necessary."

"Fifteen?"

"That's a long time," the warden said.

Trying to find a compromise, hoping to avoid both an incomplete execution and leaving the current on so long Kemmler would burn, the three men dickered like buyers at an auction. Ten, fifteen, twenty, twenty-five.

Having at last decided on the duration, the warden allowed Kemmler to enter the room. He arrived with his little entourage, and according to Dr. MacDonald's post-mortem report, "appeared strikingly calm and collected. In fact, his manner and appearance indicated a state of subdued elation, as if gratified at being the central figure of the occasion, his low order of intelligence evidently rendering him unable to fully appreciate the gravity of his situation."[13]

Wearing yellow trousers, a gray jacket and a black-and-white-checked bow tie, Kemmler might have been going out for a morning stroll. Even his shoes were well polished. Dressed and comporting himself like a gentleman, right with God, and at peace with himself, Kemmler now only needed the touch of electricity for the transformation to be complete.

The door closed behind him and was locked by an attendant. Though said to be calm and collected, even eager, still Kemmler hesitated before stepping further into

---

13   MacDonald, *Report*, 6.

the room. Gas and water pipes stretched along one wall. Hanging from the ceiling were two ordinary gas lighting fixtures. The walls had been painted a shade of quiet gray only a few months before and appeared fresh and clean.

Twenty-six witnesses were locked in with Kemmler now, staring, silent. But his attention was captured by the hulking wooden object that waited for him.

The warden signaled and one of the guards handed him a chair. He placed it in front of and a little to the right of the death seat, facing the semicircle of witnesses. Kemmler sat down calmly, looked around without any evidence of fear or even special interest in the event. The warden stood to the left of him, with his hand on the back of the chair.

"Gentlemen," he said. "This is William Kemmler." The prisoner gave a little bow. "I have warned him that he has got to die and if he has anything to say he will say it."

As though he had prepared the speech beforehand, Kemmler spoke with a composure and poise that surprised the witnesses. "Gentlemen, I wish you all good luck. I believe I am going to a good place and I am ready to go. I want only to say that a great deal has been said about me that isn't true. I am bad enough. It's cruel to make me out worse."[14]

A picture of compliance, praised afterward for his courage, "manliness," and character, Kemmler had become the unspotted lamb. It was even said that the electric chair had saved Kemmler from himself.

"All religious offerings seem imbued with the idea

---

14   *Buffalo Express*, 7 Aug. 1890.

Kemmler's execution, inaccurately depicted by a period newspaper.

of sacrifice, as a necessary condition of purgation/renewal, and the mythic motif of new life purchased with the death of the old," writes Frederick Turner.[15] The American cult of progress, represented that morning by doctors, lawyers, politicians, and ministers, was no less undergirded by these impulses than any other religion. Kemmler's death does not fit neatly into any one symbolic model of sacrifice. It draws from many: public burning, totem sacrifice, ordeal by fire, communion meal.

Though Nietzsche's *Thus Spake Zarathustra* was little known in the United States at this time, and certainly was unlikely to have been read by the men involved in Kemmler's killing, the underlying motives that drive the book—the perpetual need for overcoming, the relentless striving toward the next step in human development—were very much in the air. A quest for transcendence is and was at the heart of America's faith in itself. Our culture's uneasiness with the physical realm, our near-hysterical refusal to accept our creaturely nature, have played themselves out in the wars we've fought against each other and against nature itself. Transcendence is the flight from the crude visceral reality of our earthly form, from the dark, archaic imperative, from the "blind grindings of subterranean force, the low slow suck, the murk and ooze... the dehumanizing brutality of biology and geology" in Camille Paglia's words. The longing for transcendence is the dream of escape from our animal status, and as Paglia goes on to say, "Western science and aesthetics are attempts to revise

---

[15] Frederick Turner, *Beyond Geography*, 72.

this horror into imaginatively palatable form."[16]

The "spruce and natty" gentleman sitting in the high-tech death seat was a far more palatable image than that of a person strangling slowly at the end of a rope while hundreds cheered and jeered. He was by far a more preferable subject than the men who followed him, screaming, begging, biting like animals, cursing like damned souls as they were strapped into the same chair.

Concluding his little speech, Kemmler took off his jacket, folded it neatly and handed it to the warden. Durston then asked Kemmler to turn, in order to make sure the clothing had been cut properly for the electrodes. He found that the trousers were prepared correctly but not the shirttail underneath. Using a small pocket knife, he cut a triangle in the cloth to expose the skin at the base of Kemmler's spine. Meanwhile, Kemmler was readjusting his bowtie.

"Are your suspenders all right?" the warden asked, laying aside the knife.

"Yes, all right."

"Well then, Bill, you'd better sit down here."

Kemmler did as he was told. He settled in, arranged himself against the back of the chair like a king getting used to a new throne.

"Take it cool, Bill. I'm going to stay close beside you all the while 'til the end," Durston said, buckling a strap at the waist.

"I will, I'll take it cool."[17] Then anchoring his elbows

---

16 Camille Paglia, *Sexual Personae*, 6.
17 *Buffalo Express*, 7 Aug. 1890.

on the armrests of the chair, he pressed himself firmly against the back of the seat so that the electrode could make solid contact. Durston fumbled nervously as he put the straps on and pulled them tight. "Take your time, take your time," Kemmler said and held his arms up to make it easier for the warden to bind his chest. The witnesses glanced at each other, murmuring their amazement at Kemmler's poise.

"It won't hurt you, Bill," the warden said. "It won't hurt you at all."

"Take your time, Mr. Durston. Don't be in a hurry." Then, as the leather mask was produced, he said, "Well, I wish everybody good luck." Pastor Houghton's eyes were by now glazed with tears. He looked over at the other witnesses as though proud of the work he'd accomplished with Kemmler.

The black leather mask was brought down over Kemmler's face.

It didn't fit snugly yet, so Kemmler encouraged the warden to pull the straps tighter. As his eyes were covered, Kemmler's voice changed subtly, a note of resignation entering it now. "Do everything right, Mr. Durston. And push that down more on top of my head." The electrode was forced against the shaved spot and Kemmler's head was pressed down between his shoulders. Durston cinched the buckles on the mask so tightly the edges of the stiff black leather cut into the prisoner's skin.

"Well, I want to do the best I can and I can't do any better than that."

"You have, Kemmler, you have," Dr. Spitzka said in a

↑ "Turning on the Lightning"
Illustration from
the *Police Gazette*.

→ Newspaper illustration
of Kemmler in the chair.

quick nervous voice. "God bless you, Kemmler."

"Very well, gentlemen," the warden said, moving toward the door to the control room.[18] He disappeared briefly inside. A bell system had been rigged up to communicate with the dynamo room. Two taps on the bell and the generator was started. Two more taps and the voltage was increased. One tap meant "stop."

Far off, the engine thrummed and the belts began to turn. Dr. Fell approached the chair with a small can. Seeing the warden go through the door to the control room, he poured conductive fluid through a long tube into the electrode sponges.

The warden reappeared. "Well, everything is ready." The clock on the wall read 6:44. The two presiding medical men, Dr. Shrady and Dr. Spitzka, positioned themselves in front of the chair. A third physician, Dr. Daniels, was close by with a stopwatch.

"Goodbye William," Durston said. The word "goodbye" was the signal. Edwin Davis, hidden in the adjoining room, heard the word and pulled the switch.

Instantly, the underlying hum grew louder and more insistent. But that was noticed only as an afterthought. All attention was fixed on Kemmler. A sudden crack and he jolted into rigidity as though a long stake had been driven up through his spine. All the muscles of his body knotted, clenched, and it seemed that if he hadn't been

---

18   The switchboard was roughly 5 feet long and 3 feet wide. On the board were 36 light bulbs, and ammeter for alternating currents from 0.10 to 3.0 amperes, a Cardew voltmeter with an extra resistance coil calibrated for a range from 30 to 2,000 volts, and a jaw switch to route the current to the chair.

belted in, Kemmler would have been thrown across the room. The current poured through him in a burning torrent. Except for an index finger that closed up so tightly the nail got into the flesh of his thumb, Kemmler's body was rigid and motionless, like a hero immortalized in bronze. Beside the hum of power, the only sound in the room was the creak of the leather belts. The witnesses stared, not breathing, half expecting to see Kemmler burst out of the chair. Dr. Shrady and Dr. Spitzka edged closer, drawn like metal to a powerful magnet. A crunching sound now reached their ears: Kemmler's teeth grinding against each other.

At seventeen seconds, Dr. Spitzka said, "He's dead." MacDonald nodded his agreement.

"Enough! He's dead." The warden rushed into the control room and told Davis to shut off the current. The jaw switch was flipped to the open position and Kemmler relaxed completely, a puppet with its strings suddenly cut.

The hum of the system resumed its original quieter tone and the witnesses remembered to breathe.

The doctors gathered quickly around the chair and poked at the flesh with their fingers. Denting the skin, they watched the flesh change from red to white.

Dr. Southwick was nearly ecstatic. After taking his turn at probing the body, he turned away and addressed the witnesses. "There," he exclaimed. "There is the culmination of ten years of work and study. We live today in a higher civilization from this day."[19]

However, Dr. Lewis Balch approached the chair and

---

19   *Buffalo Express*, 7 Aug. 1890.

examined the spot on Kemmler's hand where the nail had broken through. The blood continued to flow, indicating that the heart was still beating.

Dr. Spitzka said, "Undo him. The body can be taken away now." Dr. Fell went to get his resuscitation equipment. The warden began to unfasten the chair's straps.

But Dr. Balch pointed out the wound, which continued to ooze blood. Conversation stopped. Dr. Southwick, a moment before triumphant, stared dumbly. Some of the witnesses edged closer, murmuring. Others headed for the door. But it was locked.

"Is he alive?"

Roughly two minutes had passed since the current was shut off. But there was definitely movement now in Kemmler's body. A horse sound of labored breathing filled the room.

"God! He's alive!" one of the men shouted.

Foam oozed from Kemmler's mouth. The warden drew back, hearing the gurgling noises from Kemmler's throat.

A gasp broke from the body. The chest began to heave, as though the prisoner was fighting to fill his lungs.

"He's breathing!"

"Start the current! Start the current again!" Dr. Spitzka shouted at the warden, who was fumbling to get the belts refastened.

Kemmler shook and gasped, pulling against the leather straps. Blood had pooled on the armrest and now dripped to the floor. Spittle welled from his mouth. His eyelids fluttered, as though he were struggling back to consciousness.

A period engraving of the "Execution of Kemler"
by Franklin Engraving Company of Chicago.

"For God's sake, kill him and be done with it."

After frantically yanking the headpiece back in place, Durston ran to the side room and ordered the current started up again. The bells signaled to the dynamo room and far away machinery hummed to life. Davis grasped the handle of the switch and pulled it down once more.

The current surged into Kemmler and his body jerked upright. This time the power was left on far longer than the original seventeen seconds. Now Kemmler's shoulders were pulled upward in a bizarre shrug. His lungs swelled and blood drops—from burst capillaries—formed on his face like sweat.

Adding to the horror, the second time the dynamo was started up, the belts began slipping on their pulleys. Witnesses could hear the power snapping off and on, and watched Kemmler jerk and thrash in response to the fluctuations in power. The dynamo had been placed on an ordinary wooden floor, with no precautions to make it secure. When running at top speed, it vibrated violently. Countershafts were placed on a wooden frame also resting, unanchored, on the floor. This framework was not aligned properly, so the pulleys could not run true. The belts, too, were new and had not been run enough to stretch them. When the additional current was required—with Kemmler part of the circuit much more resistance was produced—the belts came near to flying off the pulleys. Their screeching could be heard in the prison yard.

George Quinby, the district attorney who had prosecuted Kemmler, ran for the door, gagging. G.C. Bain, a newsman from Washington, fainted. Other witnesses recoiled in disgust, turning their heads away. But there was no way to hide from the smell that began to fill the room.

Two and a half minutes after the switch was thrown the second time, one of the witnesses shouted, "He's on fire!"

Smoke curled from Kemmler's head, where the hair was scorching, and could be seen rising from his back also. The smell of burning meat—sweet at first, then acrid—filled the nostrils and lungs of every man in the room.

"Cut the current," someone begged.

A blue flame flickered at the base of Kemmler's spine.

"He's dead. Shut it off!"

The smoke rose and now the smell of feces mingled with the burnt meat stench. The electrical current buzzed and crackled intermittently and the witnesses shouted for the warden to end it.

Finally, Durston gave the signal to cut the power. Kemmler slumped forward, having strained against the straps so hard he'd pulled them loose. As soon as it was over, Durston gingerly unscrewed the electrodes, trying not to burn himself on the cooked flesh. Then he led the silent and weak-kneed witnesses into the cool, dark outer chamber. Coughing could be heard in the shadows, stifled retching, low curses, and mutters of disbelief. District Attorney Quinby was the first one out of the room.

Meanwhile, Dr. Fell was wetting the sponges in the electrodes and trying to put out the fires that had spread to Kemmler's clothing. He soon had the flames out, but the smell still hung heavily in the air.

The straps were undone and the mask removed. Kemmler's eyes were only half closed. Dr. Spitzka pulled the lids open and shown a light in. The pupils didn't contract. No breath came now from the nostrils or mouth.

The body was left for several hours, giving it enough time to cool so that it could be the safely removed from the chair. Between the natural rigor mortis and the cooking effect of the electrocution (Dr. MacDonald's official report said that the flesh had been baked like "overdone beef"), Kemmler was stiff and hard. Removing him from the chair, the attendants found that he had rigidified into a sitting position and had to be carried to the autopsy table in this posture.

THE MEDICAL RECORD
A Semi-Monthly Journal of Medicine and Surgery.
August 1890

By George Shrady

Heretofore the proudest claim of science had been to save or at least prolong human life, and insure for its possessor the greatest enjoyment of its many bounties. In this instance it had been plainly diverted from its course under a paradoxical plea of high humanity. And yet men of science have lent their best efforts in this direction to humor the whims of a few cranks and "world-betterers " who imagined they could make legal murder a fine art and enforce into it an element of sentimentality which might rob it of its atrocity. While we allow that electricity had been a success as far as the killing is concerned, we must admit that we have gained little if anything over the ordinary method of execution by hanging.

It becomes a serious question if humanity is not paying too dear a price for instantaneous demolition. The awe and mystery of death are intensified a thousand-fold in anticipation of what this instrument of subtle power may do compared with either the noose of the rope, the grip of the garotterer, the smart of the knout, the bore of the bullet, or the chop of the ax. And yet to harness lightning and bolt it through the human body is thought to be one of the advances of the nineteenth century.

Chapter Eight
AFTERMATH

In the 1890s the medical standard for determining death was the ability of the body to produce heat. But as Kemmler had been cooked, this method was of questionable use. Kemmler lay, still folded in a sitting position, on the autopsy table that had been wheeled in and set up in front of the death chair.

While the doctors waited for the body to cool—none of them wanted to be accused of killing Kemmler with the scalpel—they grumbled and made comments about the botching of the execution. Dr. Fell had been denied permission to use his resuscitation device. Dr. Shrady told the others that he would have used a hypodermic on Kemmler after the first seventeen-second blast of current, in hopes of bringing him back.

Accusations and recriminations circulated freely. Why was the dynamo set up so far from the death chamber? Why did no one know the exact voltage that killed Kemmler? Why had the electrodes come loose? More than one observer mentioned that Kemmler's calm and "manliness" had made the process much less difficult than it might have been. A bigger man, especially one terrified of dying in the chair, might have struggled and easily torn the electrodes loose. District Attorney Quinby—who was deemed the closest to Kemmler's size—had expressed his fear that the contacts might come undone as the straps were fitted on him the night before.

At 9:57, roughly three hours after the execution, the table was wheeled beneath the window and the autopsy begun. Dr. MacDonald was the chief physician, but five others participated: E.C. Spitzka, George Shrady, William T. Jenkins (deputy coroner of New York), Clayton Daniels and George Fell. The body was divided up like loot, Doctors Daniels and Fell of the Buffalo contingent demanding an equal portion. Spitzka and Daniels shared the brain and spinal cord. Fell examined a specimen of the blood with a microscope and Shrady took notes in shorthand.

Rigor mortis was quite pronounced, especially in the jaw, neck, and upper chest. Discoloration showed in most of the surfaces of the body, especially on the forehead and chin where the straps had bitten into the skin. On the small of the back was a five-inch burn, with four concentric circles of greenish-brown charred flesh. When the chest cavity was opened, the blood was found to be particularly dark and fluid. The lungs were removed, tested, and found to be in a "marked emphysematous condition."[1] The heart was cut out and weighed. Spleen, gall bladder, stomach, kidneys, and bladder were all removed and examined. Confirming traditional stories about

---

1   MacDonald, *Report* 9-13.

executed men, there was a copious discharge of semen. This was examined under a microscope and found to contain a "large quantity of dead spermatozoa."

The doctors next moved on to the head. The scalp, incised and peeled back from the skull, was in a "desiccated condition" where it had made contact with the electrode. Upon removal of the skull cap, carbonized blood was found. "On the internal aspect of the calvarium the meningeal vessels in the dura and in their contents appeared to be black and carbonized. The carbonized vessels were so brittle that their ends were torn off with the calvarium and presented a broken, crummy appearance." Daniels described the blood in the brain as having an appearance similar to charcoal. Probing deeper into the brain, they found that the tissue there was still warm. Inserting a thermometer, a temperature of 97° was found. Portions of the brain and spinal column were preserved by the doctors for later examination.

Blood samples were taken, and Daniels, speaking with a *Buffalo Express* reporter the next day remarked on the change in the blood's consistency. He showed two vials of blood, and pointed out that it was oddly watery and would not coagulate.

After describing the autopsy,[2] Daniels said, "Different

---

[2] Though the autopsy was thorough, though a great deal of attention was given to the minute details of the execution, almost nothing of scientific value was gleaned from Kemmler's death. The current levels and length of contact were sufficiently different from those in accidental deaths that the finding of the doctors shed no light on how electricity kills. It was not until 1899 that ventricular fibrillation was shown to be the most common cause of death in electrocution.

doctors took away various organs of the body for special examination and study. Dr. Jenkins and myself fixed up the body as best we could after the operation, and left it on the dissection table. I suppose it was buried according to law, with quicklime, in the prison yard."[3]

In fact, what remained of Kemmler lay on the table for another twenty-four hours while Warden Durston decided what to do with it. The law demanded that the remains be buried within the prison grounds, but there was no cemetery inside Auburn's walls. The next night, after receiving legal advice, Durston had Kemmler buried in the potter's field of Auburn, Fort Hill Cemetery, on the grounds of a Cayuga Indian village. To avoid prying reporters, Kemmler's remains, such as they were, were taken out of the prison after midnight by wagon. Working by lantern light, prison laborers dug a grave and poured in fifteen bushels of quicklime to destroy the body.

So ended Kemmler's earthly existence. Yet to the men who created the electric chair, his value as a symbolic sacrificial victim continued. The odd use of food metaphors (MacDonald said the flesh was like "overdone beef," and another witness stated: "The muscles were carbonized. I have eaten beef cooked by electricity for six and a half minutes and well done, and this looked like it.")[4] indicate that something more than mere propitiation to some ill-defined god of progress was at work here. Newspapers commented on Kemmler's "oxlike submission." The doctors bottled his blood like a holy relic. And at the bizarre

---

3   *Buffalo Express*, 8 Aug. 1890.
4   *Buffalo Express*, 8 Aug. 1890.

group-autopsy, six learned men gathered around the table with scalpels and saws and forceps, vying for the best parts, like children squabbling for the drumstick on a Thanksgiving turkey. All of this makes Kemmler's death seem uncannily like a totem meal where members of a clan kill and eat the beast that defines them as a social group. Kemmler became by his death a receptacle of the electrical numen: awful, abhorred, and venerated at the same time. The men gathered in that dimly lit basement were members of a tightly knit brotherhood, a "clan" defined by its devotion to progress. By participating in the totem meal, they were able to partake of and share among themselves electricity's "Godlike power" and "Heavenly might," as *Scientific American* described it.[5]

"The victim who is sacrificed becomes a holy offering to the gods, to nature, to fate," Ernest Becker states in *The Denial of Death*. "The community gets more life by means of the victim's death, and so the victim has the privilege of serving the world in the highest possible way by means of his own sacrificial death."[6]

On the day Kemmler was killed, the *New York Times* said, "The situation would be full of horror to most men. To him, it is not, because of his strange nature. He is incapable of thinking and feeling as other men feel."[7]

The day after the killing, in the *New York Times* lead story, the word "sacrifice" appears in the first line, the killing being a "sacrifice to the whims and theories" of a

---

5   *Scientific American*, 27 Sept. 1890.
6   Ernest Becker, *The Denial of Death*, 138.
7   *New York Times*, 6 Aug. 1890.

"coterie of cranks and politicians."[8]

Kemmler's death was the last in a long chain of offertory killings. Hundreds of animals ad been sacrificed on the "altar of science." And as the time came closer for real human killing, the animals used were of a higher order. First cats and dogs, then horses, calves, and a group of orangutans were burned to death in secret before the governor of New York. Though there are no primary references to substantiate this legend, numerous secondary sources mention the use of primates for the final tests of the electric chair. Even though calves and large dogs approximated the weight of a man, what was needed was a creature almost human. So, according to folklore and reported in many books and articles, Edison's organization secretly bought a family of orangutans from southeast Asia and electrocuted them before the governor. However, the technology was apparently not perfected yet and instead of being shocked instantly to death, the first victim caught fire—fur blazing, skin peeling back as the animal helplessly screamed and thrashed. The image is grotesque, ridiculous perhaps, but it is an apt condensation of the emotions that surround the chair. It might not be too far a reach to say that by killing the orangutan, the technologists were in fact destroying the lower nature of mankind, symbolically killing the missing link so that humankind could be free to progress upward. The truth or falsity of the tale is not significant. What is important is that the popular imagination needed to create and maintain this image. As our crude animal nature shrieks

---

[8] *New York Times*, 7 Aug. 1890.

and burns, transcendence becomes possible because our connection to the biological past is severed.

Progress is based on the continual flight from our dual animal/spiritual nature. Transcendence of human reality is driven by the knowledge that no matter how progressive, humane, elevated we become, we still—each and every one of us—dies a brute biological death. Otto Rank carries this idea further: "The death fear of the ego is lessened by the killing, the sacrifice, of the other; through the death of the other, one buys oneself free from the penalty of dying, of being killed."[9]

Certainly Dr. A.P. Southwick would not have thought in these terms, but his statement to the press after leaving the prison carries echoes of the sentiment: "I tell you this is a grand thing, and is destined to become the system of legal death throughout the world." Overjoyed that his dreams had come true, Southwick said, "This is the grandest success of the age. After the execution I turned to Warden Durston, congratulated him, and said that I was one of the happiest men in the state of New York."[10]

With Kemmler's death, Southwick had gained himself a portion of immortality.

Though they had seen Kemmler writhing, burning, gasping for breath, the warden and the officiating doctors were unanimous in their claim that his death had been painless and pure. They all stated that Kemmler had lost consciousness instantly, and though there had

---

9   Otto Rank, *Will Therapy and Truth and Reality*, 130.
10 *New York Times*, 7 Aug. 1890.

been some "discoloration of the skin," the prisoner had felt nothing at all.

The newspapers—local, national and international—however, were vocal in their disagreement. "It is obvious that Kemmler did not die a painless death nor did he die instantly," the *Buffalo Express* reported the next day.[11]

"It would be impossible to imagine a more revolting exhibition," was the way the *London Times* put it.

"The scene can be better described as a disgrace to humanity. It will send a thrill of indignation throughout the civilized world. We cannot believe the Americans will allow the Electric Execution Act to stand," said the *London Standard*.

The *Chronicle* (London) stated that the events in Auburn were like something found in the "darkest chambers of the Inquisition." Closer to home, the *Troy Press* had this response: "After years of preparation, after scientists and experts had busied themselves in perfecting the apparatus that would insure instant death, the great state of New York is advertised to the world as having tortured one of its prisoners to death in a most horrible manner. Inflicting upon him agony that cannot be paralleled except by reference to the torture chamber of the dark ages."

Other papers used the phrases "awful experiment" (*New York Sun*), "Kemmler horror" (*New York Herald*), "scientific failure, without essential value to either science or civilization" (*Syracuse Standard*), "grim and repulsive act" (*Albany Express*).

---

11   The following responses from the press are compiled in *Buffalo Express*, 8 Aug. 1890.

Dr. B.W. Richardson, writing in *Scientific American* asserted that "Capital murder was never more thoroughly discredited in the late method of killing the unhappy culprit Kemmler by electrical discharge. What occurred was really worse than was prognosticated. The shocks administered were intense surface shocks, attended with extreme local action, but not affecting directly or immediately the respiratory centers. The man was really killed by a clumsy stun, for which a dexterous blow from a pole ax would have been an expeditious substitute."[12]

Again and again, newsmen and editorialists called into question the electric chair's humaneness and effectiveness. "Are these the results of the humanitarianism of the nineteenth century as embodied in the legislation of the enlightened State of New York?" the *Rochester Herald* asked. "We hope in the name of humanity and for the sake of the morals as well as the sensibilities of the people of this state that Kemmler's death will be the last, as it was the first," by electrocution.

"The people of New York State are not barbarians or cannibals. They do not want to put anybody, even the most desperate criminal, to death by means of torture. They do not fancy the smell of burning human flesh," the *New York Press* declared.

The *Evening World* painted a more grisly picture of the "subtle fires" that "broiled the wretched Kemmler to extinction. The acrid, sickening stench of burning flesh smites the nostrils of the lookers-on. A thin gray smoke curls about the rigid head. Yes: the current has won. This

---

12   *Scientific American*, 27 Sept. 1890.

is unconditional surrender. The smoking tenement tells of the evicted soul. Kemmler is dead! Ay, and the fair sweet mercy of electric death should die with him. Better, infinitely better, the one quick wrench of the neck-encircling hemp than this passage through the tortures of Hell to the relief of death."

Other newspapers called into question the motivation of the responsible parties. "Apparently the man died in agony, by slow torture," the *New York World* asserted. "The effect upon the witnesses was sickening. The effect upon the public is still more shocking, chiefly because of the attempt to do this judicial killing by torture in secret, and to conceal the facts, whatever they might be, from the public in whose name and by whose authority the killing was done."

The *Brooklyn Times* declared, "The pseudo-scientific enthusiasts who have advocated the substitution of electricity for hanging in the execution of criminals probably feel a lot better today... they have at last obtained a victim and enjoyed the task of experimenting on him to their hearts' content."

George Westinghouse, who had kept a low profile in the days leading up to the killing, was at last induced to make a statement. "It has been a brutal affair. They could have done better with an ax. My predictions have been verified. The public will lay the blame where it belongs and it will not be on us. I regard the manner of the killing as a complete vindication of our claims."[13]

And indeed there was plenty of blame to go around.

---

13   *New York Times*, 18 Aug. 1890.

Literally before Kemmler's body was cold, the doctors and scientists involved were pointing fingers at each other. Warden Durston and Dr. Spitzka took the brunt of it, but all involved in the affair found themselves tainted, scurrying away from responsibility for what the *New York Times* called "an awful spectacle."[14] Dr. Daniels accused Dr. Spitzka of botching the execution, having the current shut off too soon. Two days later, Spitzka sent a letter to newspapers attacking his erstwhile colleague, claiming that Daniels had "intentionally lied and otherwise stated the direct contrary of what he knew to be the truth," and accusing Daniels of pointing the finger to further his own career.

Asked by a reporter for a response, Daniels said, "This venom he shows is entirely unwarranted. It's a case of the underdog howling the most. The electrocution was not managed as it should have been, and I suppose under the spur of adverse criticism, Dr. Spitzka had to let off some spleen on somebody." Daniels then said that Spitzka had come to the execution with "several of his New York friends with the intention of freezing out the Buffalo delegation, but as the electrical execution movement originated at this end of the state, the Buffalo doctors felt they had a right to participate in the affair."[15]

Thomas Edison weighed into the verbal battle too, siding with the other nonmedical people in his claim that the technology was perfect. "The fault lies wholly

---

14   *New York Times*, 7 Aug. 1890.
15   The Spitzka-Daniels feud is reported in *Buffalo Express*, 8 Aug. 1890.

Alfred P. Southwick, M.D.

with the doctors," he said, claiming that the burning and Kemmler's "apparent" sufferings were due to misplacement of the electrodes. "The way to execute a criminal is to send the current through his body from one arm of the death chair to the other. The arms, hands and fingers are full of blood, which is a good conductor of electricity."[16]

Resolute in his denials of responsibility, Durston, too, placed the blame on the physicians, maintaining that they were accountable for the current being stopped too soon.

Spitzka's actual position regarding the execution is murky. In a letter to the *New York Times*, he calls himself "an avowed opponent" of electrocution, and claims that he had gone to Auburn "solely to make the examination of the brain and not as an assistant executioner." He was quoted shortly after the execution as saying, "I've seen hangings that were immeasurably more brutal, but I've never seen anything so awful. I believe this will be the first and last execution of the kind."[17] Yet three months later, at a meeting of the Society of Medical Jurisprudence, he said, "The execution of Kemmler was a more decent and dignified execution of the law than any I have ever seen."[18] He then went on to blame the uproar attending Kemmler's death on men who'd never seen an execution and were therefore unprepared for the sights and sounds they experienced that morning in Auburn. Oddly though, he voiced the opinion that the guillotine

---

16 *Buffalo Express*, 8 Aug. 1890.
17 Cited in Frederick Drimmer, *Until You Are Dead*, 15.
18 *New York Times*, 11 Nov. 1890.

was the best method of capital punishment.

Except for the sniping between the Buffalo and New York medical factions, the doctors were unanimous in their accusation that any problems related to Kemmler's death were the fault of the prison management and technology used. In his report to the governor, Dr. MacDonald—though he deems the Kemmler killing "a step in the direction of a higher civilization"—places the blame for mishaps on the arrangement of the death apparatus. MacDonald was not alone in pointing to the problems of having the controls for the chair in a separate room and the dynamo 1,000 feet distant. "Unfortunately, the voltmeter and other appliances for determining the strength of the current were not located in the execution room, hence none of the official witnesses could know precisely what the voltage was at the moment when the current was applied."[19] Dr. Fell echoed this claim that the warden and executioner were to blame.

Ethical questions were raised regarding the participation of physicians in such affairs. But generally, the conflict of interest for medical people was not seen as a problem. Their interpretation of the Hippocratic standard, "First do no harm," apparently allowed what they thought of as a "humane" taking of life. Dr. Richardson asked in *Scientific American* whether doctors "ought to lend themselves, under any circumstances to that loathsome act of playing public executioner,"[20] but his voice was barely heard in the clamor of physicians who pressed

---

19   MacDonald, *Report*, 15.
20   *Scientific American*, 27 Sept. 1890.

for further experimentation with the electric chair.

"If the design had been to cause failure at the critical moment, it is difficult to imagine a more effective arrangement to that end than existed when Kemmler took his seat in the death chair."[21] Questions about Durston's motivation in rearranging the entire set-up—after everything had been found satisfactory in tests—cast a great deal of suspicion on the warden. When he gave the signal to throw the switch he did not know the voltage being produced by the dynamo. And it was learned later that the thirty-six bulbs on the testing board were all burning when the current was routed to Kemmler, further diminishing the chair's killing power. It was reported by a "tall man with dark hair and a mustache who was said to be one of the three men operating the electrical apparatus in the secret chamber"[22] that the voltage during the first blast was between 1,000 and 1,500, and during the second between 1,500 and 2,000. Yet other witnesses were told upon leaving the chamber that the voltage had been between 700 and 1,300, far below the amount recommended for killing a man.

Spitzka stated later that the voltmeter was not working at all during the execution and that it had not been expected to work. He estimated the pressure at 700 volts. Though he did not have any hard evidence, he asserted later that it was quite plausible that the Westinghouse faction had been behind the botching of the execution. "The interests of the company who manufactured the

---

21   *New York Times,* 18 Aug. 1890.
22   *Buffalo Express,* 8 Aug. 1890.

dynamos would certainly be advanced by the defects in the machinery. They failed to kill electrical execution in the courts, but the last resort was not there. Their ends would be served quite as efficiently if this execution was a botch, as it largely was, and would consequently meet with public disapproval and condemnation, such as would demand the repeal of the law."[23]

The *New York Times* reported that, after the execution, the second person to leave the prison had been one of the men in the control room. He had gone directly across the street to the makeshift Western Union office and sent a telegram: "Execution an awful botch. Kemmler literally roasted to death." The message was addressed to the Westinghouse company.

However, the public condemnation and private finger-pointing did nothing to deter the continued use of the electric chair. Though there was some talk in the New York legislature of rescinding the law, the chair continued to be used with increasing frequency. The next electrocution was on July 7, 1891, when four men were killed at Sing Sing.[24]

The law forbade the publishing of any details of the execution. After the uproar that followed Kemmler's death, the state decided it would tighten its control. No members of the press were allowed to see the multiple execution. The "experts" who attended that morning were required to sign a statement affirming that they would

---

23   *New York Times*, 7 Aug. 1890.
24   James Slocum, Joseph Wood, Schichiok Jugigo, and Harris Smiler.

never discuss what they saw in Sing Sing's death chamber. Warden Brown took the secrecy issue so seriously that guards armed with carbines were posted to keep reporters away. Dr. Southwick was there to again witness his brainchild. In Dr. MacDonald's report, he states that all four men were dead within six minutes of entering the room.

Martin Loppy, a convicted wife-murderer, was "Governor Hill's sixth victim."[25] Again the warden made sure security was tight. Putting armed guards on the public highway in front of the prison, he ordered them to shoot any newsman who dared cross his "dead line." And again he attempted to swear the witnesses to secrecy. However, in this case more than one witness was willing to talk. At noon, when the prisoners were eating lunch, Loppy was brought to the death chamber and strapped into the chair. This time the electrodes were attached to the head and calf. It took four tries, four blasts of 1,700 volt electricity, to kill Loppy. He struggled, fought, groaned and writhed, and each time the current was shut off, a faint heartbeat still could be heard. On the fourth application of current, his left eyeball broke and the aqueous fluid ran down his cheek. As the smell of burnt flesh filled the room, witnesses turned away in nausea. Dr. MacDonald deemed the execution a "perfect success."[26]

Attempting to eliminate the burning and prolonged thrashing of the prisoner, a new electrode system was used for the seventh man to die in the chair. On February

---

25 *New York Times*, 8 Dec. 1891.
26 *New York Times*, 8 Dec. 1891.

8, 1892, Charles MacElvaine was killed at Sing Sing using the system originally recommended by Thomas Edison and Harold Brown. For this execution, the chair was altered slightly so that the prisoner's arms slanted downward, bound securely to the oaken slats. The hands were submerged in jars containing potash solution. In case the new method was found wanting, the calf and head electrodes were also attached to the prisoner. The current ran for 50 seconds across MacElvaine's arms and chest. When the power was cut, he was still quite alive. Apparently Edison had miscalculated the resistance of the human body. Giving up on the new system, Dr. MacDonald (who referred to all the executions as "experiments") had the backup electrodes connected and turned on. Again came the familiar smell of burning hair and flesh. Edison's system was never tried again.

A few months later in Auburn, Joseph Tice was more successfully killed with the standard head and lower leg electrodes. William Taylor was executed on July 27, 1893, also at Auburn. In this case, the first blast went as planned, but when the current was turned on again, nothing happened. Taylor was unconscious, but still very much alive. Edwin Davis worked the lever frantically, the warden shouted orders to his men, but no current would flow. Taylor was taken out of the chair, injected with morphine and given chloroform to keep him unconscious and out of pain, until the chair could be repaired. Almost an hour later, the limp body was strapped in a second time and killed. Again there was a public outcry for New York to give up the chair.

Eventually though, the medical men and scientists perfected the technology. In 1893, ten prisoners were killed. The next year the chair took only two lives, but over the following decades New York's electrical executioners[27] averaged about eleven a year. By 1905, the chair had taken more than 100 lives. The peak was in 1912, when New York executed twenty-two people. Other states came to adopt the electric chair, Ohio being the next in 1896. Massachusetts followed in 1898, then came New Jersey (1907), Virginia (1908), North Carolina (1909), and Kentucky (1910). Eventually twenty-five states, plus the District of Columbia, gave up the scaffold for the electric chair.

---

[27] From 1890-1914, Edwin F. Davis, 240 executions; 1914-26, John Hurlburt, 140 deaths; 1926-39, Robert G. Elliott, 387 deaths; 1939-53, Joseph Francel. These men also worked for other states; their total executions are not for New York only.

The New-York Times.

JULY 20, 1887

BUFFALO WAKES UP PREPARING TO UTILIZE THE GREAT WATER POWER OF THE NIAGARA

For years there has been more or less talk of utilizing the great water power of the Niagara River, which has been going to waste at the doors of Buffalo, and many have been the plans presented for harnessing the great current and making it do work for mankind. Some of the members of the Buffalo Business Men's Association have lately taken hold of the matter, and from the present outlook it seems that steps will be taken in the near future to have the waters of the Niagara River pay tribute to Buffalo and its manufactories before passing over the great cataract below.

Chapter 9

APOTHEOSIS

Westinghouse lost the battles, but not the war. It still remained to be seen whether AC or DC would triumph. And the last battlefield was only twenty miles from the place where William Kemmler killed Tillie Ziegler.

Beginning in the mid-1880s, attempts were made to harness the immense power of Niagara Falls. Two major problems stood in the way: the speed of the river's flow was far too great for water wheels and the prime site for the use of the falls' power—the Goat Island area—became off-limits in 1885 when the New York State Niagara Reservation was established to protect the region's natural beauty. The advantages of the falls, however, more than made up for the drawbacks. The evenness of the river's flow (four Great Lakes drained through this passage), the enormous drop, and sheer volume of water made taming the falls irresistible to engineers and financiers.[1] Not just the natural attributes of Niagara made

---

[1] Tesla wrote in his autobiography: "I was fascinated [as a youth] by the descriptions of Niagara Falls I had perused, and pictured

it attractive, but its location, too. In 1890, roughly one fifth of the population of the United States lived within 400 miles of the falls. And when Toronto and Montréal were also considered, the possibility of using the falls became even more seductive.

In 1885, a canal was dug to divert water away from the park preserve. And the next year, Thomas Evershed proposed building a tunnel two and a half miles long, which would feed water past 238 wheels. But financial backing was not forthcoming until 1889, when a group of New York bankers (among them J. Pierpont Morgan and William K. Vanderbilt) committed themselves to the long-term development of Niagara Falls power, creating the Cataract Construction Company. In 1890, they bought out the Niagara Falls Power Company and assumed its options for land use and right-of-way for the discharge tunnel. Seeking the most lucrative and up-to-date plans for mastering the falls, they requested proposals from a number of engineers and inventors.

Edison submitted a proposal that would have built a main tunnel, water turbines, a belt system for transferring the power to the dynamos far above on the surface, and transmission of electrical power via direct current to Buffalo. Westinghouse too submitted a plan, recommending that the power be transferred to the generators using compressed air.

None of the proposals was thought practical and the

> in my imagination a big wheel run by the Falls. I told my uncle that I would go to America and carry out this scheme. Thirty years later I saw my ideas carried out at Niagara and marveled at the unfathomable mystery of the mind." (48)

Cataract Company decided on further study. Edward Dean Adams was sent to Europe to consult with the leading scientists and engineers there. In June of 1890 the International Niagara Commission was established with Lord Kelvin as the chairman. The commission sponsored a competition to gather the best plans for taming Niagara. A Westinghouse engineer, L.B. Stillwell, asked his employer if they should enter the contest. Westinghouse, suspicious that this was just a way to get valuable technical information cheaply (the prize money was $20,000) refused to submit a plan. "When the Niagara people are ready to do business, we shall make them a proposal."[2] In August of that year alternating current's extraordinary power was demonstrated to the world when William Kemmler was killed. Two months later, construction began on the main Niagara tunnel. In the spring of the following year, 1891, the commission awarded its prizes, but deemed none of the plans suitable for Niagara Falls. Two important decisions were made though: that the generators would need to be on the surface, driven by vertical shafts from below, and that the means of long-distance transport of power would definitely be electrical.

Also in 1891 a Swiss company successfully transmitted alternating current over 116 miles from a waterfall on the Neckar river to Frankfurt, Germany. Edward Dean Adams consulted with this firm, which was using Tesla polyphase technology.

The commission, and especially Lord Kelvin,

---

2   Qtd. in Passer, 285 n 16.

remained adamant that direct current, not alternating current, would be the choice for transmitting Niagara Falls power. However, between 1891 in 1893, a number of successful AC installations began to erode support for DC. In Telluride, a small mining town in Colorado, Westinghouse constructed his first commercial application of Tesla motors and generators, supplying power for mining camps. And early in 1893, in Pomona, California, Westinghouse built a hydroelectric plant and a 10,000 volt, thirty-five-mile-long transmission line.

Both Westinghouse and Edison were determined to win the Niagara Falls contract; with it would come millions of dollars in profit and a great deal of esteem from the engineering profession. One stepping stone toward this goal would be the Columbian Exposition in Chicago. Commemorating the four hundredth anniversary of "Columbus taking possession" (according to the carved words on a plaque below his statue), this fair would be the biggest of its kind, a showcase for the latest technological advances, a "massive scale model of the 20th century as dreamed by the professional experts of an emerging world power."[3] Just as the 1876 Philadelphia fair had been a perfect opportunity for Edison to show off his latest inventions, so the Chicago Exposition would serve Westinghouse as an excellent stage from which he could demonstrate his technological prowess.

Accepting that he would lose money in the short term, Westinghouse underbid Edison, and the contract

---

3   James Koehnline, "Been So Long at the Fair," *Moorish Science Monitor*, 12 Apr. 1993.

to electrify the fair was awarded to him in 1892. He immediately contacted Tesla in order to plan the massive undertaking. Reluctantly, Tesla put aside the work he was doing to help his supporter and one-time employer. Though the alternating current apparatus that they would show at Chicago would be state of the art, Tesla's scientific imagination had already gone far beyond this. The exposition, however, was too important an opportunity for him to forgo.

On May 1, 1893, President Grover Cleveland—who had come a long way from his days as sheriff and hangman of Buffalo—formally opened the fair. As tens of thousands of spectators cheered and waved white handkerchieves, he took the podium to give the opening speech. Praising the "stupendous results of American enterprise and activity... the magnificent evidences of American skill and intelligence," he declared that the exposition displayed for all the world to see, "the unparalleled advancement and wonderful accomplishments of a young nation, and present[ed] the triumphs of a vigorous, self-reliant, and independent people."[4]

Although electricity had been installed in the White House in 1891, until Cleveland stood on the stage in Chicago, no president had been allowed to touch the control switches. On May 1, 1893, a meeting occurred which was no less consequential than the driving of the golden spike that linked East and West. Political power and technological power made a momentous first contact.

Cleveland placed his finger on the gold-and-ivory

---

4   *New York Times,* 2 May 1893.

Victor telegraph key that would with one push start the machinery for the entire fair. The device, symbol of American push-button ingenuity, forerunner of all American exercises in instantaneous power, rested on a pedestal upholstered in blue-and-gold plush. On the sides, the dates 1492-1893 shown in bright silver.

Cleveland smiled and pressed the key. A choir broke into the Hallelujah Chorus and the vast machinery of the fair awoke, driven by AC generators and AC motors. Gears, shafts, belts, lights, engines, and electric fountains sprang to life. At the Centennial Fair, the Bell telephone had been a mere toy. And there were only a half dozen arc lamps exhibited there as scientific curiosities. At Chicago, 17 years later, there were 90,000 incandescent bulbs and 5,000 arc lamps. If anything defines the Chicago Exhibition of 1893, it is the presence, the total victory, of electricity as a definer of American genius and mastery over nature.

The fair attracted 25 million visitors (close to one third the population of the United States at the time) in the first six months. Over 200 buildings were spread out on 666 acres along the shores of Lake Michigan. Canals were built and mock lagoons too, where gondolas glided and little steamboats puttered. Frederick Law Olmsted, architect of Central Park, contributed design. Most of the huge edifices—the smallest major building covered five and half acres—had the appearance of gleaming marble. They were, however, made of wood framing coated with a mixture of plaster and hemp fiber, sprayed with white paint from compressed-air guns. "The whole

Fair was laid out as a gigantic illustration of prevailing ideas of Social Darwinism and eugenics. On the Midway Plaisance, the long strip leading up to the White City, the Department of Ethnology arranged the pavilions of the "lesser" nations in order, from the most "savage and primitive" at the far end, to the most "civilized," near the gates of the Great White City. In the middle stood the Queen of the Midway, George Ferris's enormous wheel.[5]

Visitors came from around the world to see the first zipper, pay telephones, typewriters, calculating machines, Edison's Kinetograph (a "great novelty" later known as motion pictures), concrete paving, the debut of Cracker Jack and Aunt Jemima, the massive Krupp 120-ton siege cannon, which would twenty years later pound the fortresses of Belgium and France, the largest searchlight ever made (with a capacity of 180,000,000 candle power). And over all this presided "the official Deity, Progress, and attendant seraphim such as Science, Wealth, Power, and Civilization."[6]

The electrical building, where Edison, Westinghouse, Western Electric, and other manufacturers had their displays, opened formally on June 1 and the main event was the unveiling of Edison's Tower of Light, built by General Electric. Standing a hundred feet tall, the massive column represented "the highest achievement of the incandescent lamp." The column was topped by a replica of the Edison incandescent bulb, eight feet tall, made from small crystals of cut glass on a metal frame. More

---

5   Koehnline, "Been So Long at the Fair."
6   Koehnline, "Been so Long at the Fair."

than 30,000 of these man-made jewels were used in the construction of the column, and wired so that they would flash in different colors in time with music.

Westinghouse, too, had a display in the electrical pavilion. Under a huge mural of Columbus—his face outlined by bulbs—Westinghouse displayed his most advanced technology. And the centerpiece of the show was Nikola Tesla, demonstrating his amazing inventions.

With Westinghouse's triumph in Chicago, there was no doubt who would receive the contract to harness Niagara Falls. In January of 1893, the Cataract officials had visited the Westinghouse plant in Pittsburgh and the General Electric factories in Schenectady and Lynn to determine whether they would be able to actually build the equipment called for in the proposal. Two months later, both companies formally submitted their complete plans. In May of 1893, while the world flocked to Chicago to see the wonders Westinghouse and Tesla had created, the Cataract Company made its final decision to use alternating current. A half year later, the Cataract engineers recommended to their board of directors that Westinghouse be given the contract.

However, due to rapid expansion, Westinghouse was in a weak financial position. And because of the power of Morgan and other Cataract brokers, a compromise was worked out between the two warring factions. Westinghouse was given the contract for the first three generators and the switch gear in the powerhouse. The contract for the transformers, transmission lines, and substation equipment went to General Electric. Ten years after the first electricity was produced at Niagara, the two powerhouses contained ten Westinghouse generators and eleven made by G.E., which had agreed to a cross-licensing arrangement in order to use the Tesla patents. G.E. became, in effect, a partner with Westinghouse, both free now to produce complete lines of AC equipment. Forming what might be called a duopoly, G.E. and Westinghouse froze all smaller manufacturers out of the market until the Sherman Antitrust Act of 1911 broke up their hegemony in the field.

The Great War of the Currents had ended in a draw, both sides compelled to compromise by financial realities. However, there was no doubt that the work of Nikola Tesla had triumphed completely over all competing technologies. Using Tesla patents, the first generators went into service at Niagara Falls in August of 1895. In November of the next year the first power was transmitted to Buffalo, for use in streetcars.

On January 27, 1897, Tesla was invited by Edward Dean Adams to speak at a banquet celebrating the success of the power project. He took the podium and was met with deafening applause. People got to their feet, clapping and waving napkins. Three minutes passed before he was able to speak.

These years were the pinnacle of Tesla's career and the high point of his standing with the public. The *New York Times* stated that his was the undisputed honor of harnessing Niagara. The Order of the Eagle was conferred on him by the Prince of Montenegro, his native country. The American Institute of Electrical Engineers (AIEE) awarded him its Elliott-Cresson medal for his work with high frequency electricity. And even Lord Kelvin, originally opposed to the use of AC at Niagara, now declared that Tesla had contributed more to electrical science than any man before him.

Tesla's fame would fade; poverty, accidents, and mental illness would plague him. But during the last years of the nineteenth century, his genius remained unquestioned. In the Westinghouse exhibit at the Chicago Fair Tesla was the main draw, the paragon of progress.

He had a separate structure built at the exhibit (curtains, columns, Victorian gingerbread) and day after day would amaze his audience there with AC's power. On a velvet-covered table, copper balls, discs, and eggs were made to spin rapidly, then reverse direction at Tesla's command. Bars of iron, tin, and lead would melt when subjected to his high frequency coils. But certainly the high point of the demonstrations came when he turned himself into a figure of living fire. Thomas C. Martin describes the spectacle this way:

> Tesla has been seen receiving through his hands currents of more than 200,000 volts, vibrating at a million times per second, and manifesting themselves in dazzling streams of light... Tesla's body and clothing have continued for some time to emit fine glimmers or halos of splintered light. In fact, an actual flame is produced by the agitation of electrically charged molecules, and the curious spectacle can be seen of puissant, white, ethereal flames that do not consume

anything, bursting from the ends of an induction coil as though it were the bush on holy ground. With such vibrations as can be maintained by a potential of 3,000,000 volts, Mr. Tesla expects someday to envelop himself in a complete sheet of lambent fire that will leave him quite on injured.[7]

In the huge cathedral of electricity, under the fatherly gaze of Christopher Columbus (patron saint of progress), in a specially built chamber shrine, Tesla achieved his apotheosis. Transformed by fire that did not burn, he was filled with electricity's near supernatural power. Not only were the sheets of "cold fire" that coursed over his body harmless, but, he believed, actually therapeutic. They refreshed the mind and cleansed the body, Tesla claimed. Engulfed in electric fire, he rose—in his view—to the next step in human evolution.

In Sing Sing, Dannemora, Auburn, prisoners waited in cages to be strapped into a chair, killed, and quickly forgotten. But Tesla stood before awed crowds, a tall gaunt man who played with the basic constituents of nature, like a shaman or priest. Celebrated as the greatest electrical genius of his century, he was transformed by the same "godlike power," which, in secret basement chambers, cooked prisoners to death, nameless sacrificial animals.

---

[7] Thomas C. Martin, "A New Edison on the Horizon," *The Review of Reviews*, Mar. 1894: 355.

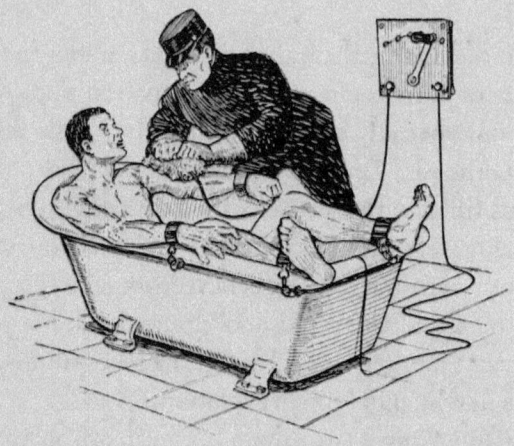

An illustration from *Crime and Criminals* by the Prison Reform League, Los Angeles: Prison Reform League Publishing Company (1910). It is titled "The Hummingbird" and goes on to state: "Chained in a metal tank the victim is tortured with electricity until his muscles chord and he faints from pain."

Bibliography

Becker, Ernest. *The Denial of Death.* New York: Free Press, 1973.

Berkson, Larry C. *The Concept of Cruel and Unusual Punishment.* Lexington, Maine: Heath, 1975.

Bernstein, Theodore. "A Great Success." *I.E.E.E. Spectrum* Feb. 1973: 54-58.

Bishop, George V. *Executions: The Legal Ways of Death.* Los Angeles: Sherbourne, 1965.

Brown, Harold P. *The Comparative Danger to Life of the Alternating and Continuous Currents.* N.p., 1889.

___. "The New Instrument of Execution." *North American Review* Nov. 1889: 586-93.

Canby, Edward T. *A History of Electricity.* New York: Hawthorn, 1963.

Chambers, Robert. *Vestiges of the Natural History of Creation.* London: Hull House, 1844.

Cheney, Margaret. *Tesla: Man Out Of Time.* New York: Dorset, 1989.

Commission to Investigate and Report The Most Humane and Practical Method of Carrying into Effect the Sentence of Death in Capital Cases. *Report.* Albany: Argus, 1888.

Conot, Robert. *A Streak of Luck.* New York: Seaview, 1979.

Clark, Ronald W. *Edison: The Man Who Made the Future.* New York: Putnam, 1977.

Cranston, Sylvia. *HPB: The Extraordinary Life and Influence Of Helena Blavatsky.* New York: Putnam, 1993.

Culianu, Ioan. *Eros and Magic in the Renaissance.* Chicago: U of Chicago P, 1987.

Dickson, W.K.L., and Antonia Dickson. *The Life and Inventions of Thomas Alva Edison.* New York: Crowell, 1894.

The Dream City. St. Louis: Thompson, 1893.

Drimmer, Frederick. *Until You Are Dead*. Secaucus, N.J.: Carol, 1990.

Edison Archives, Edison National Historical Site, West Orange, N.J.

Edison, Thomas Alva. *The Diary and Sundry Observations*. New York: Philosophical, 1948.

Eliade, Mircea. *The Forge and the Crucible*. Chicago: U of Chicago P, 1978.

Elliott, Robert G. *Agent of Death*. New York: Dutton, 1940.

Forbes, B.C. "Edison Working on How to Communicate with the Next World." *American Magazine*. Oct. 1920: 10+.

Foucault, Michel. *Discipline and Punish*. New York: Pantheon, 1977.

Franklin, H. Bruce. *War Stars: The Superweapon and the American Imagination*. New York: Oxford UP, 1988.

Gerry, Elbridge. "Capital Punishment by Electricity." *North American Review*. Sept. 1989: 321-25.

Girard, René. *Violence and the Sacred*. Baltimore: Johns Hopkins UP, 1977.

Glimpses of the World's Fair. Chicago: Laird and Lee, 1893.

Hill, David B. *Public Papers of David B. Hill, Governor*. Albany: Argus, 1886.

Hughes, Thomas P. "Harold P. Brown and the Executioner's Current: An Incident in the AC-DC Controversy." *Publications in the Humanities*. 70: 142-65.

Josephson, Matthew. *Edison: A Biography*. New York: McGraw, 1959.

Koehnline, James. "Been So Long at the Fair." *Moorish Science Monitor*. 12 Apr. 1993: 16-23.

Lehman, Susan. "A Matter of Engineering." *Atlantic Monthly*. Feb. 1990: 26-29.

Lewis, Orlando. *The Development of American Prisons and Prison Customs, 1776-1845*. N.p., 1922.

Leupp, Francis E. *George Westinghouse: His Life and Achievements*. Boston: Little, 1919.

MacDonald, Carlos F. *Report of Carlos F. MacDonald, M.D., on the Execution by Electricity of William Kemmler, alias John Hart*. Albany: Argus, 1890.

Martin, Thomas C. "A New Edison on the Horizon." *Review of Reviews*. Mar. 1894: 355.

___. *The Inventions, Researches and Writings of Nikola Tesla*. New York: *The Electrical Engineer*, 1894.

McGurrin, James. *Bourke Cockran: A Free Lance in American Politics*. New York: Scribner, 1948.

McLendon, James. *Deathwork*. Philadelphia: Lippincott, 1977.

Millard, Andre. *Edison and the Business of Innovation*. Baltimore: Johns Hopkins UP, 1990.

Nerney, Mary C. *Thomas A. Edison: A Modern Olympian*. New York: Smith and Haas, 1934.

New York Supplement. vol. 7. St. Paul: West, 1890.

Northeastern Reporter. vol. 24. St. Paul: West, 1890.

O'Neill, John J. *Prodigal Genius*. New York: Ives Washburn, 1944.

Paglia, Camille. *Sexual Personae*. New York: Vintage, 1991.

Passer, Harold C. *The Electrical Manufacturers: 1875-1900*. New York: Arno, 1972.

Prison Discipline Society. *First Report*. N.p., 1826.

Prison Discipline Society. *Fourth Report*. N.p., 1829.

Rank, Otto. *Will Therapy and Truth and Reality*. New York: Knopf, 1936.

"A Restless Philanthropist." *Electrical Engineer*. Feb. 1889: 74.

Richardson, B.W. "The Execution by Electricity." *Scientific American*. 27 Sept. 1890: 200.

Seifer, Marc. *Nikola Tesla: The Man Who Harnessed Niagara Falls.* Kingston, R.I.: Metascience, 1991.

Serviss, Garret P. *Edison's Conquest of Mars. New York Evening Journal*, 1898.

Supreme Court Reporter. vol. 10. St. Paul: West, 1890.

Tesla, Nikola. *My Inventions: The Autobiography of Nikola Tesla.* Ed. Ben Johnson. Williston, Vt.: Hart, 1982.

Turner, Frederick. *Beyond Geography: The Western Spirit Against The Wilderness.* New York: Viking, 1980.

Vogel, M., E. Patton, P. Redding. *America's Crossroads: Buffalo's Canal Street/Dante Place.* Buffalo: Heritage, 1993.

Wachhorst, Wyn. *Thomas Alva Edison: An American Myth.* Cambridge, Mass.: MIT P, 1982.

A Warning. N.p.: Edison Electric Light Company, 1888.

Weisberg, Jacob. "This Is Your Death." *New Republic.* July 1991: 23-27.

Westinghouse, George. *Safety of the Alternating System of Electrical Distribution.* N.p.: George Westinghouse, 1889.

Name Index

Adams, Edward Dean 150, 253, 259
Aldini, Giovanni 22
Alger, Horatio 77
Allison, Dr. H.E. 208
Andrews, William S. 130
Anthony, William A. 114
Argue, Dr. Henry 208

Bain, George C. 208, 227
Balch, Dr. Lewis 208, 224, 225
Barnes. C.F. 198, 199, 204, 210
Batchelor, Charles 108, 109, 143
Becker, Ernest 235
Becker, Tracy C. 175, 181, 182, 208
Bell, Alexander Graham 93, 109
Blackman, Dr. 66
Blackwell, A.B. 162
Blavatsky, Helena 89, 90
Bleyer, Doctor 145
Brown, Alfred K. 113
Brown, Harold P. 132–181, 186, 192, 203, 248
Brown, Warden William R. 247
Burns, Patrick 26
Burroughs, Edgar Rice 80

Chambers, Robert 23
Chandler, Professor 137
Cheney, Margaret 121, 124, 129
Churchill, Winston 177
Cleveland, Grover 34, 255
Cockran, W. Bourke 166, 175–189, 194, 195
Coffin, C. A. 159
Columbus, Christopher 261
Conot, Robert 70, 78

Conway, Michael 208
Cranston, Sylvia 90
Croker, Richard 176, 177
Culiano, Ioan 83, 84, 88, 89
Czolgosz, Leon 203

Daniels, Dr. C.W. 208, 223, 232, 233, 241
Davis, Edwin F. 161, 192, 200, 207, 223, 224, 227, 248, 249
Day, Judge 175, 186, 187
DiBella, John "Yellow" 55–63
Dickson, W.K.L. 82, 83
Drimmer, Frederick 243
Druse, Bill 30–35
Druse, George 31, 33
Druse, Mary 31–33
Druse, Roxalana 30, 31–39
Dugan, Sgt. 67
Duncan, Professor Louis 157
Durston, Mrs. 168, 191, 202
Durston, Warden Charles F. 163, 164, 185, 186, 198, 200–204, 207, 210, 214, 220–223, 227, 228, 234, 237, 241–245
Dwight, Judge Charles 187–189
Dwight, Louis 171

Edison, Thomas Alva 27, 51, 69, 70–117, 121, 124, 129, 130–133, 135, 142–161, 166, 168, 180, 184, 185, 196, 241, 248, 252, 254, 257
Einstein, Albert 74
Eliade, Mircea 85–88

Elliott, Robert G. 249

Faraday, Michael 11
Fell, Dr. George 18, 26, 41,
    162–164, 192, 194, 208,
    209, 223, 225, 228, 231,
    232, 244
Ferraris, Galileo 130, 131
Ferris, George 257
Fish, Frank 205, 207
Flower, Roswell P. 201
Forbes, B.C. 92
Ford, Henry 69, 81, 91
Forster, Thomas 22
Foucault, Michel 167, 168
Fowler, Dr. Joseph 208
Francel, Joseph 249
Francis, Willie 12
Fuller, Chief Justice Melville 194

Galvani, Luigi 21, 22
Garfield, President James A. 202
Gates, Charles 31
Gates, Frank 31, 32, 33
Gaulard, Lucien 122
Gerry, Elbridge 41–43, 144, 153,
    154, 159, 183, 184
Gibbs, John Dixon 122
Gilroy, Thomas 176
Girard, René 39
Goethe 105
Grant, Hugh 176
Gray, Justice 190, 191
Guillotine, Dr. Joseph-Ignace 161
Guiteau, Charles 202

Haight, Mrs. 34
Hale, Matthew 41, 42
Hankinson, Agent 140
Harrison, R.H. 25

Hastings, F. S. 135, 144
Hatch, Charles S. 175
Hearst, William Randolph 79
Hill, Governor David 33–35, 40,
    53, 133, 146, 201, 204, 247
Hillman, Harold 12
Hoover, Herbert 70
Hort, Emma 54, 59–64
Houdini, Harry 91
Houghton, Rev. Dr. O.A. 191,
    202, 205, 213, 221
Hurlburt, John 249

Irish, George 208

Jenkins, Dr. J. M. 208
Jenkins, Dr. William T. 208, 232,
    234
Jenkins, Oliver 208
Josephson, Matthew 92, 94, 143
Jugigo, Schichiok 246

Kelly, John 176
Kelvin, Lord 253, 260
Kemmler, William aka "Billy Hort"
    16, 17, 54–68, 161–168,
    174, 175, 185–217,
    220–231, 234, 235, 238,
    241–246, 251, 253
Kennelly, Arthur 138–141, 144,
    150, 166, 177
King, Asa 65, 66
Kligo. Frank 10
Koehnline, James 254, 257

Lathrop, Superintendent Austin
    98, 156, 163, 165
Laudy, Professor Louis 162, 163
Lehman, Susan 12, 15
Leutcher, Fred A. 11, 12, 15

Lewis, Orlando 172
Loppy, Martin 247

MacDonald, Dr. Carlos F. 156, 162, 164, 208, 216, 224, 229, 232, 234, 244, 247, 248
MacElvaine, Charles 248
Mack, Frank 208
MacMahon, Dennis 59, 60
Malone, Thomas 65
Martin, Thomas C. 114, 260, 261
McMichael, Dr. 18
McMillan, Daniel 37
McNamee, Graham 70
McNaughton, Daniel 191, 199, 207
Millard, Andre 129
Miskell, John 160
Morgan, J. Pierpont 117, 121, 252, 258
Moses, Dr. Otto 159

Nelson, Dr. W. T. 208
Nietzsche, Friedrich 219
Noble, John 181
Nollet, Abbé 19, 20

O'Brien, Justice 190
Olcott, H.S. 89
O'Neill, John 107, 111, 116, 118

Paglia, Camille 219, 220
Park, Dr. 18, 66
Passer, Harold 150, 151
Peck, Charles 113
Pentecost, Hugh O. 158
Petersen, Dr. Fredrick 182, 183
Petersen, Fredrick 135, 138, 143, 146

Pope, Franklin 177, 179, 180
Porter, Ida 58
Powell, Dr. 35–37

Quinby, George T. 180, 227, 228, 232

Rank, Otto 237
Reese, Bert 91
Reid, Mrs. 54, 57, 65
Richardson, Dr. B.W. 239, 244
Rockwell. Dr. 184

Serviss, Garret P. 79, 80
Shelley, Mary 21, 22
Sherman, Roger 193, 194
Shrady, Dr. George 208, 223, 224, 230–232
Sickmon, C.W. 190
Slocum, James 246
Smiler, Harris 246
Smith, T. Carpenter 166
Snyder, Ruth 2
Southwick, Dr. Alfred P. 23, 24, 37, 41, 42, 164, 192, 208, 211, 215, 224, 225, 237, 242, 247
Spitzka, Dr. Edward A. 202
Spitzka, Dr. Edward C. 202, 208, 215, 221–225, 229, 232, 241, 243, 245
Stanley, William 122
Steinmetz, Charles Proteus 74
Stillwell, L.B. 253
Szigety, Anital 105, 113

Tafero, Jesse Joseph 7–9
Taylor, William 248
Tesla, Nikola 97–131, 146, 251–255, 258–261

Thompson, Elihu 51, 123
Tice, Joseph 248
Turner, Frederick 219
Twain, Mark 117

Vail, J.H. 130
Vanderbilt, William K. 252
Vann, Judge 52
Veiling, Joseph C. 208, 213

Wachhorst, Wyn 72, 78
Wallace, Judge 193
Weisberg, Jacob 10–13
Wells, H.G. 79
Wemple, Bill 191
Westinghouse, George 122–134, 146–153, 157, 161, 165, 168, 175, 177, 195, 240, 252–258, 260
Wheeler, Schuyler 177
Wood, Joseph 246

Yates, Rev. Horatio 191, 202, 205, 208, 213
Yellow. *See* DiBella, John "Yellow"

Ziegler, Fred 56, 60
Ziegler, Matilda "Tillie" 54–67, 251

Other titles you may enjoy from
UNDERWORLD AMUSEMENTS
WWW.UnderworldAmusements.com

The Radical Book Shop of Chicago:
In Which a Disaffected Preacher, His Blind Anarchist Wife, and Their Precocious Daughters Create an Important Hub of Literary, Bohemian, and Revolutionary Culture in Progressive-Era Chicago
by Kevin I. Slaughter,
with appendix by Lillian Undell
130 pages, 6x9", $16.95
ISBN: 978-1943687282

New York is Hell
Thinking and Drinking in the Beautiful Beast
by Benjamin DeCasseres
introduction by Peggy Nadramia
6x9, 360 pages, $18.95
ISBN: 978-0988553606

Confessions of a Failed Egoist
And Other Essays
by Trevor Blake
5.25x8, 139 pages, $9.95
ISBN: 978-0988553651

The Philosophical Writings of Edgar Saltus
Containing the books "The Philosophy of Disenchantment" and "The Anatomy of Negation"
Foreword by Chip Smith
6x9, 366 pages, $18.95
ISBN: 978-0988553644

The Sorceries and Scandals of Satan
by Henry M. Tichenor
foreword by R. Merciless
6x9, 176 pages, $15.95
ISBN: 978-0983031406

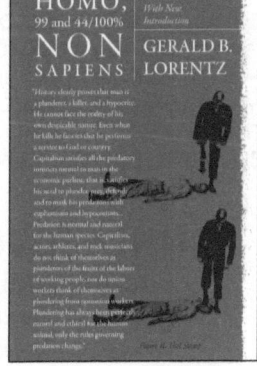

Homo, 99 and 44/100% Nonsapeins
Revised with a new Introduction
by Gerald B. Lorentz
6x9, 424 pages, $18.95
ISBN: 978-0988553637

STANDARD FREETHOUGHT WORKS

CONFESSIONS OF A FAILED EGOIST—*Trevor Blake* $10
ELBERT HUBBARD'S THE PHILISTINE—*Bruce A. White* $16
HOMO 99 AND 44/100 NONSAPIENS—*Gerald B. Lorentz* $18
MIGHT IS RIGHT: THE AUTHORITATIVE EDITION—*Ragnar Redbeard* . . . $20
MIGHT IS RIGHT: 1927 FACSIMILE EDITION—*Ragnar Redbeard* $16
THE OCCULT TECHNOLOGY OF POWER—*The Transcriber* $8
THE PHILOSOPHICAL WRITINGS OF EDGAR SALTUS—*Edgar Saltus* . . . $18
THE RADICAL BOOK SHOP OF CHICAGO—*Kevin I. Slaughter* $16
THE RED SECT—*Enzo Martucci* $16
RIVAL CAESARS: A ROMANCE...—*Ragnar Redbeard* $20
THE SATANIC SCRIPTURES—*Peter H. Gilmore* $17
SORCERIES AND SCANDALS OF SATAN—*Henry M. Tichenor* $15
THIS UGLY CIVILIZATION—*Ralph Borsodi* $20

BENJAMIN DeCASSERES SERIES:
ANATHEMA! LITANIES OF NEGATION $10
FANTASIA IMPROMPTU & FINIS $16
FULMINATIONS: CAUSTIC, COSMIC, CAPRICIOUS $16
IMP: THE POETRY OF BENJAMIN DeCASSERES $15
NEW YORK IS HELL: THINKING AND DRINKING IN THE BEAUTIFUL BEAST $18
SPINOZA: LIBERATOR OF GOD AND MAN & AGAINST THE RABBIS . . . $15
THE BOY OF BETHLEHEM—Bio DeCasseres (Hardbound) $23
THE SUBLIME BOY—*Walter DeCasseres* $7

THE PORTABLE L.A.ROLLINS SERIES:
THE MYTH OF NATURAL RIGHTS $15
LUCIFER'S LEXICON $15
OUTLAW HISTORY $15
DISJECTA MEMBRA . . . . (coming soon)

PAMPHLETS

BOVARYSM: THE ART-PHILOSOPHY OF JULES DE GAULTIER—*Wilmot E. Ellis* $4
IMMORALITY AS A PHILOSOPHIC PRINCIPLE—*Paul Carus* $5
MAX STIRNER AND THE PHILOSOPHY OF THE INDIVIDUAL—*Leo Markun* . $8
MAN-EATING AND MAN-SACRIFICING—*Anon.* $3
THE NIETZSCHE MOVEMENT IN ENGLAND—*Oscar Levy* $2
PRIMITIVES: POEMS AND WOODCUTS—*Max Weber* $6

UNDERWORLD AMUSEMENTS
444 MARYLAND AVE. #7940   ESSEX, MD 21221

For postage add $4 for the first item, $1 for each additional.

Or visit WWW.UNDERWORLDAMUSEMENTS.COM

www.ingramcontent.com/pod-product-compliance
Lightning Source LLC
Chambersburg PA
CBHW060408130526
44592CB00046B/1019